KING

D1128288

DATE DUE / DATE DE RETOUR

MAY 2 5 2002		
FEB 2 1 2012		
MAR 0 1 2016		
AUG 3 0 2016		

UNDERSTANDING
COSMOLOGY

FROM THE EDITORS OF *SCIENTIFIC AMERICAN*

Compiled and with Introductions by

Sandy Fritz

Foreword by Donald Goldsmith

A Byron Preiss Book

WARNER BOOKS

An AOL Time Warner Company

The essays in this book first appeared in the pages of *Scientific
American,* as follows: "How Cosmology Became a Science," August
1992; "The Evolution of the Universe," 1999 special issue,
Magnificent Cosmos; "The Self-Reproducing Inflationary Universe,"
Magnificent Cosmos; "Echoes from the Big Bang," January 2001;
"Inflation in a Low-Density Universe," January 1999; "The Quintes-
sential Universe," January 2001; "Cosmological Antigravity," January
1999; "Quantum Cosmology and the Creation of the Universe,"
December 1991; "The Fate of Life in the Universe," November 1999.

Warner Books, Inc., 1271 Avenue of the Americas,
New York, NY 10020

Visit our Web site at www.twbookmark.com.

 An AOL Time Warner Company

Printed in the United States of America
First Printing: March 2002
10 9 8 7 6 5 4 3 2 1

ISBN: 0-446-67873-2
Library of Congress Control Number: 2001093909

Cover design by J. Vita
Book design by Gilda Hannah

Contents

I some fits and starts, the
ry seemed to best accommoda
growing observations. Tiny
tuations found throughout t
ic microwave background ech
erse. These in turn were one of t
y hallmark proofs of the bi
rs soon followed. Most mode
ological thought frames its
big bang theory. When its

Foreword: THE GOLDEN AGE OF COSMOLOGY

Donald Goldsmith

Cosmology has entered a golden era, an epoch when new generations of astronomical instruments have flung wide the doors that once blocked our view of the universe. Streaming inward through these observational opportunities, a flood of new information has set our old concept of the cosmos on its ear, shocking the conservatives and stimulating a new generation of astronomers and cosmologists who cherish the belief that we may eventually uncover the secrets of the universe.

What is this information that has so markedly changed our understanding of the cosmos? High-altitude, balloon-borne, and satellite observing platforms now allow us to detect, to record, and to analyze cosmic radiation within spectral domains previously banned from our scrutiny by the Earth's atmosphere. In addition, tremendous improvements in ground-based telescopes, still the largest instruments devoted to studying the universe, provide us with a host of high-resolution images of objects detected at ever greater distances, now well in excess of 12 billion light years. With these enormous distances, we travel backward in time more than four-fifths of the way to the big bang, the moment when all space and matter crowded together at near-infinite density.

The most startling new results in cosmology have emerged from observations that combine the power of ground-based optical and radio telescopes with the additional information carried in other regions of the electromagnetic spectrum but accessible only by avoiding most or all of the Earth's atmosphere. The outstanding example of this observational synergy appears in the recent discovery, now confirmed almost beyond reasonable doubt, that the expansion of the universe, rather than slowing down as the result of the mutual gravitational attraction between each part of the cosmos and all the other parts, is actually accelerating!

This amazing conclusion now rests on two separate and complementary pillars of observational evidence, from observations of exploding stars billions of light years beyond our galaxy and from studies of the cosmic background radiation produced in even more distant regions of the universe. Exploding stars of a particular kind, called Type Ia supernovae, can be observed at tremendous distances with the Hubble Space Telescope and the Keck Telescopes in Hawaii. Because the Type Ia supernovae all reach approximately the same maximum luminosity, their peak apparent brightnesses reveal their distances from us. By comparing these distances with the speeds at which these exploding stars are moving away from us, available from observations of the Doppler effect in the spectra of the light from the galaxies within which they appear, astronomers have found that the most distant of these exploding stars are somewhat farther away than astronomers anticipated. Something has caused the universe to expand more rapidly than accepted cosmic models called for. Is this really so? And if it is, what can that "something" be?

Observations of the Cosmic Microwave Background radiation (CMB) imply that the acceleration of the expanding universe, first revealed by the Type Ia supernovae, can and must be a fact of nature. The CMB arose a few hundred thousand years after the big bang, when the universe had thinned itself by expansion to the point that the radiation that filled it ceased to interact with matter. Since that time, this radiation, which we now detect as the CMB, has been affected only by the expansion of the universe, so it provides a snapshot of the early history of the universe, faded by the passage of time but still laden with information.

By observing the CMB's detailed properties, astronomers can deduce how much matter and energy must exist, on the average, in every cubic centimeter of space. These quantities determine how the cosmic expansion will change with time. Like the observations of Type Ia supernovae, studies of the CMB show that the universe teems with what cosmologists now call "dark energy," a peculiar, utterly invisible form of energy that fills all space and has the unusual property of making space itself expand. This energy drives the cosmos into an accelerating mode, producing ever more rapid expansion as time goes on. The dark energy has a noble pedigree, for it corresponds to the "cosmological constant" that Albert Einstein introduced into his basic equation to describe the evolution of the universe. Nevertheless, we know nothing at all about the dark energy, save the crucial fact that it exists! A new generation of particle physicists and cosmologists must someday provide an explanation of how empty space can be packed with energy, so that the accelerating universe becomes the ultimate free lunch, producing ever more energy as it expands into ever greater amounts of space.

The discovery of the dark energy answers one of the great cosmological questions: Will the universe expand forever, or will it someday contract, perhaps to produce a "big crunch" and a recycled cosmos? We now know, subject to the possibility that the great discoveries of the past few years are wrong, that the universe will expand forever, and at an increasing rate, so that within a few hundred billion years, the cosmos will be dark and nearly empty throughout its enormous volume. If that is so, we must count ourselves fortunate to live in an era when stars still exist in abundance, furnishing the raw materials and energy for civilizations to exist nearby, where beings with a curiosity in their origin and fate may also contemplate the forces that drive the cosmos.

UNDERSTANDING COSMOLOGY

r some fits and starts, the
ory seemed to best accommoda
growing observations. Tiny
tuations found throughout t
ic microwave background ech
verse's birth were one of t
y hallmark proofs of the bi
rs soon followed. Most mode
ological thought frames its
big bang theory. When its

Introduction

Sandy Fritz

"**Y**ou Are Here," states the caption on a poster of a spiral galaxy suspended in space-time. The arrow points to an invisible speck embedded in a curved galactic arm composed of millions of stars. When considering ourselves on a scale of this magnitude, our planet seems like a subatomic particle in an expanding matrix 15 billion light-years wide and filled with countless galaxies.

More than any other science, cosmology stretches the human mind. It challenges the mind to consider the distance a beam of light travels in a year at its constant speed of 186,000 miles per second. This is the measure of space and time called a light year, and is equivalent to six trillion miles. Then consider that one of the Milky Way's nearest neighbors, the Andromeda Galaxy, is five million light years distant from us. It challenges our common sense to understand that in the dark reaches of space-time, subatomic particles appear from nothingness, blink a physical presence for 5.3×10^{-44} seconds and then vanish—only to return again in another 55.3×10^{-44} seconds, creating an endless cycle of pulsing that may be causing the universe to expand. To learn such things, and to fit them into our scale of human events, is a humbling experience.

The men and women who contemplate nature on this scale are a special breed, and their work is a brilliant testimony to the range and flexibility of the human mind. Their ponderings have led us to examine the early moments of the universe's birth, when subatomic particles brewed in a microscopic energy cauldron. In it, all energy and all matter that would become the universe waits in potential. Then, in a hyper-instant, gravity waves and space-time inflate, and the pent up universe expands out into the budding matrix and fills it. And if our observations are correct, it is filling it still.

The accumulation and understanding of this information took time to accomplish. In many ways it was initiated by Edwin Hubble, an American cosmologist who, in 1930, woke the world to the reality that the universe is composed of millions and millions of galaxies. And these galaxies, he showed us, are traveling away from each other at near-light speeds, suggesting that the universe is expanding.

The finding took Albert Einstein, whose general theory of relativity postulated a static universe, by surprise. The findings took the whole world by surprise and set in motion an earnest start to the systematic probing of the universe for its secrets.

After some fits and starts, the big bang theory seemed to best accommodate our growing observations. Tiny temperature fluctuations found throughout the cosmic microwave background—echoes of the universe's birth—were one of the early hallmark proofs of the big bang. Others soon followed.

Most modern cosmological thought frames itself around the big bang theory. When its shortcomings challenged the theory's boundaries, the inflation theory was born. Borrowing from the field of particle physics, the inflation theory suggests that energy fields "seeded" the universe, and that the big bang expansion of energy and matter followed.

No theory is perfect, and human curiosity cannot help but

pick at the model it has created to refine it more and more. Curious questions have arisen from this refinement process. For example, what powers the expansion of the universe? Is it some kind of mysterious matter, invisible to our current technology, that urges the universe outward in defiance of the inward pull of gravity inherent in all physical matter? In "Cosmological Antigravity," by Lawrence Krauss, a variety of possible explanations are explored. Each option carries ramifications that could alter our fundamental view of the cosmos.

One possible driver of cosmic expansion may be an energy form that varies over time. "The Quintessential Universe," by Jeremiah Ostriker and Paul Sternhardt, considers this possibility, postulating that as the conditions of the universe evolved over time, so did the nature of the energy that fills it.

That changes took place over time in the evolving universe seems certain. Specific relics from the universe's earliest moments should be, in theory, present and observable, and some important findings have been coaxed from cosmic microwave background radiation. Still illusive, however, are gravity waves, which are believed to have not only expanded out into spacetime in advance of the big bang but also squeezed and stretched the raw material of the universe itself. "Echoes from the Big Bang," by Robert Caldwell and Mark Kamiankowski, considers this possibility and details their role in establishing a geometry for the universe.

How the universe expanded and what its overall shape may be is the topic of "Inflation in a Low-Density Universe," by Martin Bucher and David Spergel. What is the universe's shape: is it curved like a bubble, as predicted in some variations of the inflation theory? Bent like a potato chip? Or is the universe flat like a thin sheet of rubber? The way we view the universe has a profound impact not only on our overall conceptualization of the cosmos but also on the ultimate rise or fall of such theories that predict them.

Perhaps the most astounding insights arise when theorists

employ quantum mechanics—the study of very small particles—to frame their perception of the macro scale of the cosmos. Suddenly, the universe becomes far more dynamic than we expected: particles pop in and out of existence, history becomes probability, and an infinite number of possible scenarios play out simultaneously in galaxies that are infinite in number. "Quantum Cosmology," by Jonathan Halliwell, takes the reader into some of these realms, then catalogs the frantic attempts of cosmologists to tame the wild theories into more manageable, testable hypotheses.

Some theories tackle the universe on such an ambitious scale that finding proof for them stretches human ingenuity to new heights. The "grand unification" notion, sometimes referred to as the theory of everything, attempts to explain the four fundamental forces of nature (gravity, electromagnetism, the weak and strong atomic forces) as manifestations of a single force.

Observation and testing are the keys to confirmation and acceptance of any scientific theory, but sometimes observations dangle outside the framework, awaiting explanation. The universe seems to be expanding more quickly than can be accounted for in accepted theories, and the density of matter in the universe may be far less than anyone has anticipated.

The next-generation orbital cosmological probe, dubbed MAP, should be an important tool in answering these and other questions. Its sensitive gear will point into deep space and map the universe in the microwave spectrum. With a resolution greater than any previous microwave probe, MAP should be able to provide unprecedented detail of the cosmos, and will probably end up raising more questions than it answers.

Another next-generation tool situated on the peak of a dormant volcano in Hawaii is the Gemini North Telescope, scanning the cosmos on an infrared wavelength in conjunction with a soon to be built sister telescope, Gemini South, in Peru. The logistics of establishing such cosmological tools is, not surprisingly, a very earthly concern, littered with nuts, bolts, and the hammers of funding realities.

Within the dictates of the linear time that we experience, there is a beginning to the universe, and thus there should be an end as well. The lesson applies to the star that powers life on Earth. If it was born, it must therefore someday die. What will happen to the human species then? "The Fate of Life in the Universe" muses on the subject, and projects a scenario where human consciousness could dwell among the stars.

These questions and others are the topic of this book, culled from the pages of *Scientific American* magazine with the intent to trace the birth of these inquiries, to track their development, and to ponder some of the key questions facing cosmologists in the 21st century. Because we are made of the same subatomic particles of which all matter in the universe is comprised, we too are a part of the greater whole, linked to it with a connection as natural as a quasar, as mysterious as virtual particles.

*Few people remember the "match of the heavyweight scientists" in the 1950s and 1960s that established the big bang as the cornerstone of modern cosmology. Indeed, at first the big bang was the "challenger," squaring off against the theory known as steady state, then hailed by many cosmologists as the heavyweight champion of the universe. The contest was to be judged by an acid test derived from the doctrine set forth by Karl Popper—*a scientific hypothesis must predict phenomena that can be tested. *The steady state theory, failing to predict or account for cosmic microwave radiation, fell by the wayside.*

How Cosmology Became a Science

Stephen G. Brush

Did the universe begin, or has it always existed? Scientists long regarded this question as lying outside their concern, in the metaphysical realm of philosophers and theologians. Not until the middle of this century did physicists and astronomers begin to equip themselves with theories powerful enough and experimental techniques sensitive enough to address the issue.

Two competing cosmologies then emerged. One, popularly called the big bang, assumes that the universe evolved from initial conditions so hot and dense that only radiation and elementary particles could exist; the universe then expanded and cooled, precipitating the stars and galaxies. The opposing model offers a universe that has always existed; the dispersal of matter resulting from the observed expansion of the universe is compensated by the continuous creation of matter.

The big bang theory has prevailed, largely because of the prediction, observation and interpretation of a phenomenon known as the cosmic background radiation. This radiation, widely regarded as the afterglow of the big bang, suffuses the sky in all directions at microwave frequencies. Arno A. Penzias and Robert W. Wilson of Bell Laboratories discovered the cosmic background in 1964–65 while trying to rid their radio antenna of microwave noise. (See "Echos from the Big Bang," page 46.) The steady state model of the universe predicted no such radiation and could not plausibly account for it. Thus, for the first time, hypotheses about the origin of the cosmos had faced an empirical test that left a winner and a loser.

Rarely do theories stand or fall on the outcome of a single test. This time, however, opinion shifted almost overnight. Within a few years, most cosmologists had either adopted the big bang theory or ceased publishing in the field.

Yet no one could have appreciated the significance of the cosmic microwave background without the legacy of knowledge that many other scientists had been building throughout the century. The history of the discovery yields another kind of insight. By following the story past 1965 to see how the discovery affected the standing of rival cosmological theories, we can test competing ideas about the nature of scientific progress.

Big bang cosmology began to come into focus in the 1930s, after Edwin P. Hubble, the eminent American astronomer, showed that galaxies appear to recede from one another and that the most distant ones recede at the greatest rate. Hubble's finding implies that the universe is expanding. It was also interpreted to imply that the cosmos had once been concentrated in a very small space at a definite time. Alexander A. Friedman, a Russian physicist, and Georges Lamaître, a Belgian priest, each used Albert Einstein's general theory of relativity to describe how such an expanding universe might evolve.

Nuclear physics played a role by providing the tools with which to model the synthesis of the elements from fundamental particles. Those tools served not only George Gamow, champion of the big bang, and his colleagues Ralph A. Alpher and Robert Herman but also British cosmologist Fred Hoyle—

then at the University of Cambridge—who favored the rival steady state theory.

Vital to the theoretical work was the contribution that Einstein and Max Planck made around the turn of the century while formulating the physics of blackbody radiation. The blackbody gets its name from its idealized property of absorbing all incoming radiation and then reradiating it. This reradiated energy is distributed across the spectrum in a highly characteristic pattern, predicted by Planck. Because the primordial fireball, in its early phases, would have put energy and matter into perfect thermal equilibrium, the first radiation liberated from the cooling explosion would have to have displayed the blackbody pattern.

Still to be supplied was a precise calculation of how energetic that spectral pattern would appear today, many billions of years after the fireball began to expand and cool. What was the temperature of the radiation in space? An answer to that question could come only after scientists developed a quantitative theory of the evolution of the fireball after the big bang.

The development of this quantitative theory began with Gamow, a Russian-born physicist who had made his reputation by explaining radioactive decay. In the 1930s he came to the U.S., teaching first at George Washington University and then at the University of Colorado. At George Washington, he concentrated on the astrophysical and cosmological aspects of nuclear reactions—above all, the mechanisms by which the first elements had been synthesized.

Gamow looked for his answer at both ends of the cosmic scale. In the early 1930s astronomers showed that most stars were composed predominantly of hydrogen and helium. It was reasonable to assume that hydrogen was the first element to form because its nucleus contains but a single proton and that helium—the next heaviest element, whose nucleus contains two protons and two neutrons—was the first "higher" element formed by the fusion of hydrogen. But protons will fuse only if some force overcomes the immense electrostatic repulsion between them. This process seemed to require so much heat and pressure that only a primordial event or the interior of a star could have provided the right conditions.

The reigning theory of the nuclear physics of stars, which remains for the most part valid today, had been developed in 1938 by the German-born physicist Hans Bethe. Bethe wanted to explain how the sun shines. He did so by assuming that nuclear fusion in stellar interiors converts mass into energy. Specifically, Bethe proposed that two fusion reactions could take place in stars like the sun: one fuses protons into helium nuclei, and another adds protons to carbon nuclei to form heavier elements.

But where did the carbon originate? That question was not answered until the 1950s, when Hoyle proposed a reaction that could produce carbon from three helium nuclei under the special conditions found at the core of a star. That reaction and others needed to create heavier elements were confirmed experimentally, in a high-energy particle accelerator. By 1957 a scheme explaining how stars might have synthesized most of the elements from hydrogen and helium had been worked out. Yet the cosmic abundance of helium remained a mystery.

Gamow had already formulated a daring hypothesis that ultimately led to the solution of the helium puzzle. In his version of the big bang, Gamow suggested that the elements might have formed even before the stars came into being, in a stupendously hot and dense gas of neutrons. Some of the neutrons would then have decayed into protons and electrons— the building blocks of hydrogen. In 1948 Gamow, known for his impatience with detail as well as for his brilliance, assigned the task of developing the theory to Ralph Alpher, a graduate student at George Washington. Alpher later joined forces with Robert Herman of the Johns Hopkins University Applied

Physics Laboratory. Alpher gave Gamow's initial substance the name "ylem" from a Greek word meaning "primordial matter."

According to Gamow's theory as worked out by Alpher and Herman, larger nuclei formed in the primeval inferno when smaller ones, beginning with hydrogen, grew through the successive capture of neutrons. The process continued until the supply of free neutrons ran out, the temperature fell and the particles dispersed.

Alpher and Herman soon realized that the radiation pervading their model universe would maintain the spectrum of a blackbody source as it cooled. Moreover, they could calculate how the expansion of the universe would have attenuated this radiation and reduced its temperature. The two scientists used estimates of the present density of matter in the universe to predict the temperature of the cosmic background radiation today and derived a value of about five kelvins (degrees Celsius above absolute zero).

Astronomers did not rush to confirm the prediction, perhaps because they did not know how to pick out the cosmic background from other radiative sources or perhaps because they did not take seriously the cosmology on which the prediction was based. The original version of the big bang theory had two major drawbacks. First, it failed to explain the formation of the elements beyond helium, which has a mass number of four. Because there are no stable isotopes having mass numbers of five and eight, one cannot make heavier elements out of helium by adding neutrons one at a time. This problem could be solved only by invoking the stellar nucleosynthesis of Hoyle and his collaborators, a concept associated with the steady state theory. Indeed, the modern version of the big bang theory assumes that elements beyond helium arose only after the formation of the first generation of stars.

A second objection to a big bang universe involved the question of age. Astronomical measurements of the distances and recessional speeds of galaxies, in conjunction with Hubble's law of expansion, implied that the universe was two billion years old. Yet the rocks of the earth's surface prove that the planet is significantly older than that.

The steady state theory was conceived to resolve this apparent contradiction. One night in 1946, three young scientists in Cambridge, England—Hoyle, Hermann Bondi and Thomas Gold—went to see a ghost story film, *Dead of Night*. As Hoyle later recalled the movie, it "had four separate parts linked ingeniously together in such a way that the film became circular, its end the same as its beginning." Gold asked his friends whether the universe might be similarly constructed. In the ensuing discussion the workers sketched out a dynamic but noncyclic model of the universe that would always look the same even though it is always changing.

According to Hoyle, Bondi and Gold, the universe had no beginning. They argued that the galaxies' rushing away from us does not imply a continuous attenuation of matter: our own galaxy will never be left all alone, they said, because matter is being created continuously, at a rate just sufficient to compensate for the matter that is disappearing from the visible universe. This new matter will eventually form stars and galaxies, so that the universe will always look about the same to any observer at any time.

One might object that the creation of matter out of nothing violates the law of conservation of mass and energy. The riposte is obvious: the big bang also violates this law and does so by creating matter all at once, at the beginning of time, when the act is beyond the reach of scientific study.

Proponents of the steady state asserted that their theory was more scientific than the big bang because it postulated a process—continuous creation—that might in principle be observed. Moreover, they argued, their theory made definite predictions of a kind that astronomers could test in the near future.

In staking their model on the outcome of a small number of

observations, Bondi, Gold and other proponents of the steady state model explicitly invoked the doctrine of the late Karl Popper, an Austrian-born philosopher. Popper defines science as a discipline founded on the creation of hypotheses that predict phenomena—preferably new ones—that can be tested. If a prediction fails, the scientist abandons the hypothesis; if the hypothesis survives, the scientist does not claim to have proved it but merely to have established the hypothesis as a basis for further research.

Bondi proposed to challenge the steady state theory by comparing the universe as it is with the universe as it once was. Because the steady state theory says the universe always looks the same, it predicts that galaxies formed recently will resemble those formed long ago. If you look out into space—and thus back in time, because the speed of light is finite—and see that distant galaxies are different from nearby ones, Bondi concluded, "then the steady-state theory is stone dead." Like others writing before 1965, however, Bondi failed to mention another test of the steady state model: it does not predict a cosmic microwave background.

The theory failed the test Bondi had set it. In the 1950s and early 1960s a variety of astronomical observations showed that the universe had changed significantly over time. Martin Ryle of Cambridge counted both distant and nearby radio sources, knowing that the more distant signals had taken longer to arrive and thus reflected an earlier stage in cosmic history. Ryle concluded that there had been fewer sources in the past. Although some astronomers argued that he had not proved his case, additional supporting evidence emerged when astronomers discovered what seemed to be the oldest radiative sources—quasistellar objects, or quasars. These objects had no contemporary parallel whatsoever.

Meanwhile the awkward issue of the disparity between the age of the universe and the age of the Earth was resolved in a way that favored the big bang. In 1952, following the lead of

Walter Baade of the Mount Wilson Observatory, astronomers revised their scale of galactic distances upward by a factor of two. The estimated age of the universe therefore doubled. Later work raised it to a minimum of 10 billion years, whereas the age of the earth remained fixed at 4.5 billion years.

Yet many scientists, particularly in Britain, liked the simplicity of the steady state theory and so continued to cling to the concept. They pointed out that one did not have to make arbitrary assumptions about a big bang or worry about what happened before the big bang. Advocates of the steady state model also took heart from the failure of earlier attempted refutations, a record that made them suspicious of any new attacks.

As the steady staters spent ever more time explaining away the evidence accumulating against their theory, their adherence to Popper's methodology steadily became less credible. Instead they seemed to be illustrating Planck's more cynical view of science. Writing in his *Scientific Autobiography and Other Papers* (1949), the great physicist argued, "A new scientific truth does not triumph by convincing its opponents and making them see the light, but rather because its opponents eventually die, and a new generation grows up that is familiar with it."

Planck's principle, as historians of science now call it, contradicts Popper's principle by emphasizing the human element in science to the detriment of abstract logic. Just as astronomers can weigh the big bang against the steady state as a description of the universe, so may historians of science try to decide between Planck's and Popper's descriptions of science. Let us see which seems more accurate in this particular case, without undertaking to judge whether science always works in this way.

In 1959 a survey showed that a majority of astronomers rejected continuous creation, although only a third of those voting actually favored the big bang. Even Hoyle abandoned his original model and replaced it with a more complicated hypothesis. In 1964 he concluded that the high abundance of helium in the universe implied it had been "cooked" at temperatures exceeding 10^{10} kelvins. Yet Hoyle refused to abandon the idea of the continuous creation, and discovery of the cosmic microwave background provided that shock. Penzias and Wilson made the discovery by measuring the temperature of space.

One can infer a temperature of space indirectly. As Arthur Stanley Eddington pointed out in 1926, the amount of light coming from all the stars—that is, the total energy density—would be equivalent to 3.2 kelvins if converted to thermal equilibrium. But Eddington did not propose a specific procedure for testing his prediction.

At that time, even a scientist of Eddington's caliber would have found the task daunting. Obviously, ordinary thermometers would be swamped by energy coming from the sun, other celestial objects and the Earth's atmosphere. Only exceedingly sensitive instruments, tuned to wavelengths between a millimeter and a centimeter and insulated from local sources, can hope to detect the cosmic microwaves.

About 15 years after Eddington made his prescient prediction, Andrew McKellar of the Dominion Astrophysical Observatory in Canada suggested a practical way to measure what he called the effective temperature of space. McKellar, one of the first astronomers to propose that molecules as well as atoms could exist in interstellar space, suggested that the cyanogen (CN) molecule be employed as a thermometer. He noted that cyanogen emits spectral lines whose relative intensity corresponds to the number of electrons in higher-energy states—itself a function of the temperature of space; McKellar estimated that temperature to be 2.3 kelvins.

These indirect approaches could not rule out interference from local sources. To do that, one must detect the radiation itself and map it across the sky. Radar equipment developed at the Massachusetts Institute of Technology during World War II was just barely capable of sensing the cosmic background directly—for anyone who wanted to look for it.

In 1946 a group at M.I.T., led by Robert H. Dicke reported atmospheric radiation measurements taken by a microwave radiometer. The team noted that the "radiation from cosmic matter at the radiometer wavelengths" was quite sparse—less than the equivalent of 20 kelvins—but did not follow up this observation. Dicke, who subsequently moved to Princeton University, later recalled that "at the time of this measurement we were not thinking of the 'big bang' radiation but only of a possible glow emitted by the most distant galaxies in the universe."

Steven Weinberg, in his book *The First Three Minutes*, suggests two reasons why no one made a systematic search for the background radiation before 1965. First, the big bang had lost some credibility when it failed to explain the formation of elements heavier than helium, so that it did not seem important to test the theory's other predictions. In contrast, nucleosynthesis in stars—a theory linked to steady state cosmology—seemed to explain how heavy elements could have been made from hydrogen and helium, even though it did not explain how the helium had formed in the first place.

Second, Weinberg points to a breakdown of communication between theorists and experimentalists. The theorists did not realize one could observe the radiation with existing equipment, and the experimentalists did not realize the significance of their observations. From this perspective it is noteworthy that Dicke, who is both a theorist and an experimentalist, played a major role: together with P. James E. Peebles, he helped to relate a peculiar microwave noise to cosmological theory.

The most remarkable missed opportunity resulted from a misunderstanding between Gamow and Hoyle. Although each criticized the other's theory, they could still have friendly discussions. In the summer of 1956 Gamow told Hoyle that the universe must be filled with microwave radiation at a temperature of about 50 kelvins. (He arrived at this estimate on his own, after Alpher and Herman had published their prediction.)

As it happened, Hoyle was familiar with McKellar's proposal

that the temperature of space is about three kelvins. So Hoyle argued that the temperature could not be as high as Gamow claimed. But neither of them realized that if a direct measurement could confirm the three-kelvin value it would refute the steady state theory, which—as Hoyle recognized—predicts a zero temperature for space.

A different kind of communication problem—satellite relays—did lead to the discovery of the cosmic microwave background. Bell Labs wanted its satellites to convey as much information as possible at microwave frequencies, a task that required its workers to find and eliminate noise from all sources. The relay hardware, deriving from the firm's war-related work on radar, consisted of a horn-shaped receiver that Bell Labs engineers Harald T. Friis and A. C. Beck had built in 1942. Another Bell Labs engineer, Arthur B. Crawford, carried the idea much further. In 1960 he built a 20-foot horn receiver at the Crawford Hill facility near Holmdel, N.J. That reflector, originally used to receive signals bounced from a plastic balloon high in the atmosphere, became available for other purposes just in time for Penzias and Wilson.

The two investigators wanted to start a research program in radio astronomy. To prepare the highly sensitive instrument for their work, Penzias and Wilson first had to rid it of microwave noise. They failed in their first few attempts. Finally, in January 1965, Penzias heard that Peebles had a theory that might explain the origin of the stubbornly persistent signal.

Peebles was working with Dicke at Princeton, about 25 miles from the Holmdel laboratory. Dicke rejected the assumption that the universe necessarily began with the big bang. He thought it more likely that the universe went through phases of expansion and contraction. At the end of each contraction, he conjectured, all matter would pass through temperatures and densities intense enough to break down the heavier nuclei into protons and neutrons.

Thus, although Dicke's universe did not start with a big bang, each of its cycles must begin in a similar cataclysm. Moreover, Dicke's cosmology implied an initial fireball of high-temperature radiation that retains its Planckian blackbody character as it cools down, and he estimated that the present temperature of the radiation would be 45 kelvins. He evidently had forgotten his own 1946 measurement that suggested the existence of background radiation at a temperature less than 20 kelvins. Peebles made further calculations from Dicke's theory and obtained an estimate of about 10 kelvins.

Dicke and Peebles, together with two graduate students, P. G. Roll and D. T. Wilkinson, then started to construct an antenna at Princeton to measure the cosmic background radiation. Before they had a chance to get any results, Dicke received a call from Penzias, suggesting they get together to discuss the noise in the Crawford antenna, corresponding to a temperature of about 3.5 kelvins. It was soon apparent that Penzias and Wilson had already detected the radiation predicted by Dicke and Peebles and earlier by Alpher and Herman. But until the two astronomers talked to Dicke and Peebles, they did not know what they had found. The theoretical interpretation was essential to turn mere detection into true discovery. That discovery came more than a decade late because the scientific world had simply overlooked the earlier work by Gamow, Alpher and Herman.

The reports of the groups from Bell Labs and Princeton were sent to the *Astrophysical Journal* in May 1965 and appeared together in the July 1 issue. Publication unleashed a flood of articles in both the mass media and the scientific journals. Even Hoyle admitted that the steady state theory, at least in its original form, "will now have to be discarded," although he later tried to hang on to a modified version that could explain the microwave radiation. But Bondi's emphasis on the testability of the steady state theory had come back to haunt its proponents. Any attempt to twist the theory to explain the new discoveries risked being labeled as pseudoscience.

Although the press was quick to conclude that Penzias and Wilson had confirmed the big bang definitively, scientists realized that their results were limited to only a few wavelengths clustered at one end of the Planck curve. Other explanations of the background radiation, such as a combination of radio sources, could explain those data points. It was not until the mid-1970s that enough measurements at different frequencies had been made to convince the skeptics that the background radiation actually follows Planck's law.

By the late 1970s nearly all the original supporters of the steady state model had explicitly abandoned it or simply stopped publishing on the subject. A survey of American astronomers conducted at that time by Carol M. Copp of California State University at Fullerton found that a large majority supported the big bang over the steady state.

The rapid demise of the steady state theory after 1965 shows that Popper's principle, rather than Planck's, applies in this case. The discovery of the cosmic microwave background, combined with arguments about helium abundance and observations of distant radio sources and quasars, convinced most steady staters that their theory was no longer worth pursuing. It had been tried and found wanting.

Although the big bang is now accepted, some puzzles still remained unsolved. For example, the microwave background seemed too smooth. It lacked the slight variations in temperature and, by implication, in density that seemed necessary to send later gravitational clumping. Without such seeding, there would not have been sufficient time to produce the galaxies and supergalactic structures now observed.

Then, in April of 1992, George P. Smoot and his colleagues at the University of California at Berkeley and at the Lawrence

Berkeley Laboratory released evidence that may fill this gap in the big bang theory. They announced an analysis of measurements of the cosmic background radiation gathered by an orbiting observatory called the Cosmic Background Explorer (COBE). The data showed slight temperature variations in the cosmic background, just as had been expected by big bang theorists. The researchers interpret these "ripples" as fluctuations in the density of matter and energy in a very early phase of cosmic history. Such ripples may help explain how matter clumped under the force of its own gravity in time to form the stars, galaxies and larger structures of the contemporary universe.

Did the universe really begin at the big bang, or was there a previous contraction phase—a "big crunch"—that led to the high temperature and density? Will the universe continue to expand forever, or will it eventually collapse into a back hole? Does the creation of the universe involve quantum theory in a fundamental way? These ideas now dominate physical thought. That scientists consider such questions worthy of serious investigation is itself largely a consequence of the discovery of the cosmic microwave background, which transformed cosmology into an empirical science.

Once the big bang established itself as the premier theory guiding inquiry into the nature of the universe, cosmologists set about refining and testing their model. P. James E. Peebles ranks among the most influential supporters of this theory and over the decades has been in the forefront of a variety of issues and challenges confronting the big bang adherents. Here, Peebles and his collaborators outline the theory in detail and explore some of the questions that are still unsolved by big bang cosmology.

The Evolution of the Universe

P. James E. Peebles,
David N. Schramm,
Edwin L. Turner
and Richard G. Kron

At a particular instant roughly 12 to 15 billion years ago, all the matter and energy we can observe, concentrated in a region smaller than a dime, began to expand and cool at an incredibly rapid rate. By the time the temperature had dropped to 100 million times that of the Sun's core, the forces of nature assumed their present properties, and the elementary particles known as quarks roamed freely in a sea of energy. When the universe had expanded an additional 1,000 times, all the matter we can measure filled a region the size of the solar system.

At that time, the free quarks became confined in neutrons and protons. After the universe had grown by another factor of 1,000, protons and neutrons combined to form atomic nuclei, including most of the helium and deuterium present today. All of this occurred within the first minute of the expansion. Conditions were still too hot, however, for atomic nuclei to capture electrons. Neutral atoms appeared in abundance only after the expansion had continued for 300,000 years and the universe was 1,000 times smaller than it is now. The neutral atoms then began to coalesce into gas clouds, which later evolved into stars. By the time the universe had expanded to one fifth its present size, the stars had formed groups recognizable as young galaxies.

When the universe was half its present size, nuclear reactions in stars had produced most of the heavy elements from which terrestrial planets were made. Our solar system is relatively young: it formed five billion years ago, when the universe was two thirds its present size. Over time the formation of stars has consumed the supply of gas in galaxies, and hence the population of stars is waning. Fifteen billion years from now stars like our Sun will be relatively rare, making the universe a far less hospitable place for observers like us.

Our understanding of the genesis and evolution of the universe is one of the great achievements of 20th-century science. This knowledge comes from decades of innovative experiments and theories.

Our best efforts to explain this wealth of data are embodied in a theory known as the standard cosmological model or the big bang cosmology. The major claim of the theory is that in the large-scale average, the universe is expanding in a nearly homogeneous way from a dense early state. At present, there are no fundamental challenges to the big bang theory, although there are certainly unresolved issues within the theory itself.

Yet the big bang model goes only so far, and many fundamental mysteries remain. What was the universe like before it was expanding? What will happen in the distant future, when the last of the stars exhaust the supply of nuclear fuel? No one knows the answers yet.

Our universe may be viewed in many lights—by mystics, theologians, philosophers or scientists. In science we adopt the plodding route: we accept only what is tested by experiment or observation. Albert Einstein gave us the now well-tested and accepted general theory of relativity, which establishes the relations between mass, energy, space and time. Einstein showed that a homogeneous distribution of matter in space fits nicely with his theory. He assumed without discussion that the universe is static, unchanging in the large-scale average.

In 1922 the Russian theorist Alexander A. Friedman realized that Einstein's universe is unstable; the slightest perturbation would cause it to expand or contract. At that time, Vesto M. Slipher of Lowell Observatory was collecting the first evidence that galaxies are actually moving apart. Then, in 1929, the eminent astronomer Edwin P. Hubble showed that the rate a galaxy is moving away from us is roughly proportional to its distance from us.

The existence of an expanding universe implies that the cosmos has evolved from a dense concentration of matter into the present broadly spread distribution of galaxies. Fred Hoyle, an English cosmologist, was the first to call this process the big bang. Hoyle intended to disparage the theory, but the name was so catchy it gained popularity. It is somewhat misleading, however, to describe the expansion as some type of explosion of matter away from some particular point in space.

That is not the picture at all: in Einstein's universe the concept of space and the distribution of matter are intimately linked; the observed expansion of the system of galaxies reveals the unfolding of space itself. An essential feature of the theory is that the average density in space declines as the universe expands; the distribution of matter forms no observable edge. In an explosion the fastest particles move out into empty space, but in the big bang cosmology, particles uniformly fill all space. The expansion of the universe has had little influence on the size of galaxies or even clusters of galaxies that are bound by gravity; space is simply opening up between them. In this sense, the expansion is similar to a rising loaf of raisin bread. The dough is analogous to space, and the raisins, to clusters of galaxies. As the dough expands, the raisins move apart. Moreover, the speed with which any two raisins move apart is directly and positively related to the amount of dough separating them.

The evidence for the expansion of the universe has been accumulating for some 60 years. The first important clue is the redshift. A galaxy emits or absorbs some wavelengths of light more strongly than others. If the galaxy is moving away from us, these emission and absorption features are shifted to longer wavelengths—that is, they become redder as the recession velocity increases. This phenomenon is known as the redshift.

Hubble contributed to another crucial part of the picture. He counted the number of visible galaxies in different directions in the sky and found that they appear to be rather uniformly distributed. The value of Hubble's constant seemed to be the same in all directions, a necessary consequence of uniform expansion. Modern surveys confirm the fundamental tenet that the universe is homogeneous on large scales. Although maps of the distribution of the nearby galaxies display clumpiness, deeper surveys reveal considerable uniformity.

To test Hubble's Law, astronomers need to measure distances to galaxies. One method for gauging distance is to observe the apparent brightness of a galaxy. If one galaxy is four times fainter than an otherwise comparable galaxy, then it can be estimated to be twice as far away. This expectation has now been tested over the whole of the visible range of distances.

Some critics of the theory have pointed out that a galaxy that appears to be smaller and fainter might not actually be more distant. Fortunately, there is a direct indication that objects whose redshifts are larger really are more distant. The evidence comes from observations of an effect known as gravitational lensing. An object as massive and compact as a galaxy can act as a crude lens, producing a distorted, magnified image (or even many images) of any background radiation source that lies behind it. Such an object does so by bending the paths of light rays and other electromagnetic radiation. So if a galaxy sits in the line of sight between Earth and some distant object, it will bend the light rays from the object so that they are observable. The object behind the lens is always found to have a higher redshift than the lens itself, confirming the qualitative prediction of Hubble's Law.

Hubble's Law has great significance not only because it describes the expansion of the universe but also because it can be used to calculate the age of the cosmos. To be precise, the time elapsed since the big bang is a function of the present value of Hubble's constant and its rate of change. Astronomers have determined the approximate rate of the expansion, but no one has yet been able to measure the second value precisely.

Still, one can estimate this quantity from knowledge of the universe's average density. One expects that because gravity exerts a force that opposes expansion, galaxies would tend to move apart more slowly now than they did in the past. The rate of change in expansion is thus related to the gravitational pull of the universe set by its average density. If the density is that of just the visible material in and around galaxies, the age of the universe probably lies between 10 and 15 billion years. (The range allows for the uncertainty in the rate of expansion.)

Yet many researchers believe the density is greater than this minimum value. So-called dark matter would make up the difference. A strongly defended argument holds that the universe is just dense enough that in the remote future the expansion will slow almost to zero. Under this assumption, the age of the universe decreases to the range of seven to 13 billion years.

Estimates of the expansion time provide an important test for the big bang model of the universe. If the theory is correct, everything in the visible universe should be younger than the expansion time computed from Hubble's Law.

These two time scales do appear to be in at least rough concordance. For example, the oldest stars in the disk of the Milky Way galaxy are about nine billion years old—an estimate derived from the rate of cooling of white dwarf stars. The stars in the halo of the Milky Way are somewhat older, about 12 billion years—a value derived from the rate of nuclear fuel consumption in the cores of these stars. The ages of the oldest known chemical elements are also approximately 12 billion years—a number that comes from radioactive dating techniques. Workers in laboratories have derived these age estimates from atomic and nuclear physics. It is noteworthy that their results agree, at least approximately, with the age that astronomers have derived by measuring cosmic expansion.

The test is simple conceptually, but it took decades for astronomers to develop detectors sensitive enough to study distant galaxies in detail. When astronomers examine nearby galaxies that are powerful emitters of radio wavelengths, they see, at optical wavelengths, relatively round systems of stars. Distant radio galaxies, on the other hand, appear to have elongated and sometimes irregular structures. Moreover, in most distant radio galaxies, unlike the ones nearby, the distribution of light tends to be aligned with the pattern of the radio emission.

Likewise, when astronomers study the population of massive, dense clusters of galaxies, they find differences between those that are close and those far away. Distant clusters contain bluish galaxies that show evidence of ongoing star formation. Similar clusters that are nearby contain reddish galaxies in which active star formation ceased long ago. Observations made with the Hubble Space Telescope confirm that at least some of the enhanced star formation in these younger clusters may be the result of collisions between their member galaxies, a process that is much rarer in the present epoch.

So if galaxies are all moving away from one another and are evolving from earlier forms, it seems logical that they were once crowded together in some dense sea of matter and energy.

When the universe was very young and hot, radiation could not travel very far without being absorbed and emitted by some particle. This continuous exchange of energy maintained a state of thermal equilibrium; any particular region was unlikely to be much hotter or cooler than the average. When matter and energy settle to such a state, the result is a so-called thermal spectrum, where the intensity of radiation at each wavelength is a definite function of the temperature. Hence, radiation originating in the hot big bang is recognizable by its spectrum.

Astronomers have studied this radiation in great detail using the Cosmic Background Explorer (COBE) satellite and a number of rocket-launched, balloon-borne and ground-based experiments. The cosmic background radiation has two distinctive properties. First, it is nearly the same in all directions. Second, the spectrum is very close to that of an object in thermal equilibrium at 2.726 kelvins above absolute zero. To be sure, the cosmic background radiation was produced when the universe was far hotter than 2.726 kelvins, yet researchers anticipated correctly that the apparent temperature of the radiation would be low. In the 1930s Richard C. Tolman of the California Institute of Technology showed that the temperature of the cosmic background would diminish because of the universe's expansion.

The cosmic background radiation provides direct evidence that the universe did expand from a dense, hot state, for this is the condition needed to produce the radiation. In the dense, hot early universe, thermonuclear reactions produced elements heavier than hydrogen, including deuterium, helium and lithium. It is striking that the computed mix of the light elements agrees with the observed abundances. That is, all evidence indicates that the light elements were produced in the hot young universe, whereas the heavier elements appeared later, as byproducts of the thermonuclear reactions that power stars.

The theory for the origin of the light elements emerged from the burst of research that followed the end of World War II. George Gamow and graduate student Ralph A. Alpher of George Washington University and Robert Herman of the Johns Hopkins University Applied Physics Laboratory and others used nuclear physics data from the war effort to predict what kind of nuclear processes might have occurred in the early universe and what elements might have been produced.

Alpher and Herman also realized that a remnant of the original expansion would still be detectable in the existing universe.

Although significant details of this pioneering work were in error, it forged a link between nuclear physics and cosmology. The workers demonstrated that the early universe could be viewed as a type of thermonuclear reactor. As a result, physicists have now precisely calculated the abundances of light elements produced in the big bang and how those quantities have changed because of subsequent events in the interstellar medium and nuclear processes in stars.

Our grasp of the conditions that prevailed in the early universe does not translate into a full understanding of how galaxies formed. Nevertheless, we do have quite a few pieces of the puzzle. Gravity causes the growth of density fluctuations in the distribution of matter because it more strongly slows the expansion of denser regions, making them grow still denser. This process is observed in the growth of nearby clusters of galaxies, and the galaxies themselves were probably assembled by the same process on a smaller scale.

The growth of structure in the early universe was prevented by radiation pressure, but that changed when the universe had expanded to about 0.1 percent of its present size. At that point, the temperature was about 3,000 kelvins, cool enough to allow the ions and electrons to combine to form neutral hydrogen and helium atoms. The neutral matter was able to slip through the radiation and to form gas clouds that could collapse into star clusters.

A pressing challenge now is to reconcile the apparent uniformity of the early universe with the lumpy distribution of galaxies in the present universe. Astronomers know that the density of the early universe did not vary by much because they observe only slight irregularities in the cosmic background radiation. So far it has been easy to develop theories that are consistent with the available measurements, but more critical tests are in progress. In particular, different theories for galaxy formation predict quite different fluctuations in the cosmic background radiation. Measurements of such tiny fluctuations have not yet been done, but they might be accomplished in the generation of experiments now under way. It will be exciting to learn whether any of the theories of galaxy formation now under consideration survive these tests.

In following the debate on such matters of cosmology, one should bear in mind that all physical theories are approximations of reality that can fail if pushed too far. Physical science advances by incorporating earlier theories that are experimentally supported into larger, more encompassing frameworks. The big bang theory is supported by a wealth of evidence: it explains the cosmic background radiation, the abundances of light elements and the Hubble expansion. Thus, any new cosmology surely will include the big bang picture. Whatever developments the coming decades may bring, cosmology has moved from a branch of philosophy to a physical science where hypotheses meet the test of observation and experiment.

Despite the apparent sturdiness of the big bang model, and although it accurately predicts many observed phenomena detected in the cosmos, it fails to address several important questions. Russian-born cosmologist Andrei Linde calls the shortcomings "highly suspicious."

In response to these problems, Linde and others developed an idea known as the inflationary universe. It does not supersede the big bang but rather precedes it, and in doing so provides an elegant solution of myriad details unexplained by big bang cosmology. The theory of the self-reproducing inflationary universe is grounded in modern quantum mechanics and not only reflects insights from that branch of science but also accommodates the notion of an infinite number of probable universes.

The Self-Reproducing Inflationary Universe

Andrei Linde

I f my colleagues and I are right, we may soon be saying good-bye to the idea that our universe was a single fire-ball created in the big bang. We are exploring a new theory based on a notion that the universe went through a stage of inflation. During that time, the theory holds, the cosmos became exponentially large within an infinitesimal fraction of a second. At the end of this period, the universe continued its evolution according to the big bang model. As workers refined this inflationary scenario, they uncovered some surprising consequences. One of them constitutes a fundamental change in how the cosmos is seen. Recent versions of inflationary theory assert that instead of being an expanding ball of fire the universe is a huge, growing fractal. It consists of many inflating balls that produce new balls, which in turn produce more balls, ad infinitum.

Cosmologists did not arbitrarily invent this rather peculiar vision of the universe. Several workers, first in Russia and later in the United States, proposed the inflationary hypothesis that is the basis of its foundation. We did so to solve some of the complications left by the old big bang idea. In its standard form, the big bang theory maintains that the universe was born about 15 billion years ago from a cosmological singularity—a state in which the temperature and density are infinitely high. Of course, one cannot really speak in physical terms about these quantities as being infinite. One usually assumes that the current laws of physics did not apply then. They took hold only after the density of the universe dropped below the so-called Planck density, which equals about 1,094 grams per cubic centimeter.

As investigators developed the theory, they uncovered complicated problems. For example, the standard big bang theory, coupled with the modern theory of elementary particles, predicts the existence of many superheavy particles carrying magnetic charge—that is, objects that have only one magnetic pole. These magnetic monopoles would have a typical mass 1,016 times that of the proton, or about 0.00001 milligram. According to the standard big bang theory, monopoles should have emerged very early in the evolution of the universe and should now be as abundant as protons. In that case, the mean density of matter in the universe would be about 15 orders of magnitude greater than its present value, which is about 10^{-29} grams per cubic centimeter.

Questioning Standard Theory

This and other puzzles forced physicists to look more attentively at the basic assumptions underlying the standard cosmological theory. And we found many to be highly suspicious. I will review six of the most difficult. The first, and main, problem is the very existence of the big bang. One may wonder, What came before? If space-time did not exist then, how could everything appear from nothing? What arose first: The universe or the laws determining its evolution? Explaining this initial singularity—where and when it all began—still remains the most intractable problem of modern cosmology.

A second trouble spot is the flatness of space. General relativity suggests that space may be very curved, with a typical radius on the order of the Planck length, or 10^{-33} centimeters. We see, however, that our universe is just about flat on a scale of 1,028 centimeters, the radius of the observable part of the universe. This result of our observation differs from theoretical expectations by more than 60 orders of magnitude.

A similar discrepancy between theory and observations concerns the size of the universe, a third problem. Cosmological examinations show that our part of the universe contains at least 10^{88} elementary particles. But why is the universe so big? If one takes a universe of a typical initial size given by the Planck length and a typical initial density equal to the Planck density, then, using the standard big bang theory, one can calculate how many elementary particles such a universe might encompass. The answer is rather unexpected: the entire universe should only be large enough to accommodate just one elementary particle—or at most 10 of them. Obviously, something is wrong with this theory.

The fourth problem deals with the timing of the expansion. In its standard form, the big bang theory assumes that all parts of the universe began expanding simultaneously. But how could all the different parts of the universe synchronize the beginning of their expansion?

Fifth, there is the question about the distribution of matter in the universe. On the very large scale, matter has spread out with remarkable uniformity. Across more than 10 billion light-years, its distribution departs from perfect homogeneity by less than one part in 10,000. For a long time, nobody had any idea why the universe was so homogeneous. But those who do not have ideas sometimes have principles. One of the cornerstones of the standard cosmology was the "cosmological principle," which asserts that the universe must be homogeneous. This assumption, however, does not help much, because the universe incorporates important deviations from homogeneity, namely, stars, galaxies and other agglomerations of matter. Hence, we must explain why the universe is so uniform on large scales and at the same time suggest some mechanism that produces galaxies.

Finally, there is what I call the uniqueness problem. Albert Einstein captured its essence when he said, "What really interests me is whether God had any choice in the creation of the world." Indeed, slight changes in the physical constants of nature could have made the universe unfold in a completely different manner. For example, many popular theories of elementary particles assume that space-time originally had considerably more than four dimensions (three spatial and one temporal). In order to square theoretical calculations with the physical world in which we live, these models state that the extra dimensions have been "compactified," or shrunk to a small size and tucked away. But one may wonder why compactification stopped with four dimensions, not two or five. Moreover, the manner in which the other dimensions become rolled up is significant, for it determines the values of the constants of nature and the masses of particles. In some theories, compactification can occur in billions of different ways.

All these problems (and others I have not mentioned) are extremely perplexing. That is why it is encouraging that many of these puzzles can be resolved in the context of the theory of the self-reproducing, inflationary universe.

The basic features of the inflationary scenario are rooted in the physics of elementary particles. So I would like to take you on a brief excursion into this realm—in particular, to the unified theory of weak and electromagnetic interactions. Both these forces exert themselves through particles. Photons mediate the electromagnetic force; the W and Z particles are responsible for the weak force. But whereas photons are massless, the W and Z particles are extremely heavy. To unify the weak and electromagnetic interactions despite the obvious differences between photons and the W and Z particles, physicists introduced what are called scalar fields.

Although scalar fields are not the stuff of everyday life, a familiar analogue exists. That is the electrostatic potential—the voltage in a circuit is an example. Electrical fields appear only if this potential is uneven, as it is between the poles of a battery or if the potential changes in time. If the entire universe had the same electrostatic potential—say, 110 volts—then nobody would notice it; the potential would seem to be just another vacuum state. Similarly, a constant scalar field looks like a vacuum: we do not see it even if we are surrounded by it.

Scalar Fields

Scalar fields play a crucial role in cosmology as well as in particle physics. They provide the mechanism that generates the rapid inflation of the universe. Indeed, according to general relativity, the universe expands at a rate (approximately) proportional to the square root of its density. If the universe were filled by ordinary matter, then the density would rapidly decrease as the universe expanded. Thus, the expansion of the universe would rapidly slow down as density decreased. But because of the equivalence of mass and energy established by Einstein, the potential energy of the scalar field also contributes to the expansion. In certain cases, this energy decreases much more slowly than does the density of ordinary matter.

The persistence of this energy may lead to a stage of extremely rapid expansion, or inflation, of the universe. This possibility emerges even if one considers the very simplest version of the theory of a scalar field. In this version the potential energy reaches a minimum at the point where the scalar field vanishes. In this case, the larger the scalar field, the greater the potential energy. According to Einstein's theory of gravity, the energy of the scalar field must have caused the universe to expand very rapidly. The expansion slowed down when the scalar field reached the minimum of its potential energy.

One way to imagine the situation is to picture a ball rolling down the side of a large bowl. The bottom of the bowl represents the energy minimum. The position of the ball corresponds to the value of the scalar field. Of course, the equations describing the motion of the scalar field in an expanding universe are somewhat more complicated than the equations for the ball in an empty bowl. They contain an extra term corresponding to friction, or viscosity. This friction is akin to having molasses in the bowl. The viscosity of this liquid depends on the energy of the field: the higher the ball in the bowl is, the thicker the liquid will be. Therefore, if the field initially was very large, the energy dropped extremely slowly.

The sluggishness of the energy drop in the scalar field has a crucial implication in the expansion rate. The decline was so gradual that the potential energy of the scalar field remained almost constant as the universe expanded. This behavior contrasts sharply with that of ordinary matter, whose density rapidly decreases in an expanding universe. Thanks to the large energy of the scalar field, the universe continued to expand at a speed much greater than that predicted by preinflation cosmological theories. The size of the universe in this regime grew exponentially.

This stage of self-sustained, exponentially rapid inflation did not last long. Its duration could have been as short as 10^{-35} second. Once the energy of the field declined, the viscosity nearly disappeared, and inflation ended. Like the ball as it reaches the bottom of the bowl, the scalar field began to oscillate near the minimum of its potential energy. As the scalar field oscillated, it lost energy, giving it up in the form of elementary particles. These particles interacted with one another and eventually settled down to some equilibrium temperature. From this time on, the standard big bang theory can describe the evolution of the universe.

The main difference between inflationary theory and the old cosmology becomes clear when one calculates the size of the universe at the end of inflation. Even if the universe at the beginning of inflation was as small as 10^{-33} centimeter, after 10^{-35} second of inflation this domain acquires an unbelievable size. According to some inflationary models, this size in centimeters can equal $10^{10^{12}}$—that is, a 1 followed by a trillion zeros. These numbers depend on the models used, but in most versions, this size is many orders of magnitude greater than the size of the observable universe, or 10^{28} centimeters.

This tremendous spurt immediately solves most of the problems of the old cosmological theory. Our universe appears smooth and uniform because all inhomogeneities were stretched $10^{10^{12}}$ times. The density of primordial monopoles and other undesirable "defects" becomes exponentially diluted. The universe has become so large that we can now see just a tiny fraction of it. That is why, just like a small area on a surface of a huge inflated balloon, our part looks flat. That is why we do not need to insist that all parts of the universe began expanding simultaneously. One domain of a smallest possible size of 10^{-33} centimeter is more than enough to produce everything we see now.

An Inflationary Universe

In 1982 I introduced the so-called new inflationary universe scenario, which Andreas Albrecht and Paul J. Steinhardt of the University of Pennsylvania also later discovered. This scenario shrugged off the main problems of Guth's model. But it was still rather complicated and not very realistic.

Only a year later did I realize that inflation is a naturally emerging feature in many theories of elementary particles, including the simplest model of the scalar field discussed earlier. There is no need for quantum gravity effects, phase transitions, supercooling or even the standard assumption that the universe originally was hot. One just considers all possible kinds and values of scalar fields in the early universe and then checks to see if any of them leads to inflation. Those places where inflation does not occur remain small. Those domains where inflation takes place become exponentially large and dominate the total volume of the universe. Because the scalar fields can take arbitrary values in the early universe, I called this scenario chaotic inflation.

In many ways, chaotic inflation is so simple that it is hard to understand why the idea was not discovered sooner. I think the reason was purely psychological. The glorious successes of the big bang theory hypnotized cosmologists. We assumed that the entire universe was created at the same moment, that initially it was hot and that the scalar field from the beginning resided close to the minimum of its potential energy. Once we began relaxing these assumptions, we immediately found that inflation is not an exotic phenomenon invoked by theorists for solving their problems. It is a general regime that occurs in a wide class of theories of elementary particles. That a rapid stretching of the universe can simultaneously resolve many difficult cosmological problems may seem too good to be true. Indeed, if all inhomogeneities were stretched away, how did galaxies form? The answer is that while removing previously existing inhomogeneities, inflation at the same time made new ones.

These inhomogeneities arise from quantum effects. According to quantum mechanics, empty space is not entirely empty. The vacuum is filled with small quantum fluctuations. These fluctuations can be regarded as waves, or undulations in physical fields. The waves have all possible wavelengths and move in all directions. We cannot detect these waves because they live only briefly and are microscopic.

In the inflationary universe the vacuum structure becomes even more complicated. Inflation rapidly stretches the waves. Once their wavelengths become sufficiently large, the undulations begin to "feel" the curvature of the universe. At this moment, they stop moving because of the viscosity of the scalar field (recall that the equations describing the field contain a friction term). The first fluctuations to freeze are those that have large wavelengths. As the universe continues to expand, new fluctuations become stretched and freeze on top of other frozen waves. At this stage one cannot call these waves quantum fluctuations anymore. Most of them have extremely large wavelengths. Because these waves do not move and do not disappear, they enhance the value of the scalar field in some areas and depress it in others, thus creating inhomogeneities. These disturbances in the scalar field cause the density perturbations in the universe that are crucial for the subsequent formation of galaxies.

Testing Inflationary Theory

In addition to explaining many features of our world, inflationary theory makes several important and testable predictions. First, density perturbations produced during inflation affect the distribution of matter in the universe. They may also accompany gravitational waves. Both density perturbations and gravitational waves make their imprint on the microwave background radiation. They render the temperature of this radiation slightly different in various places in the sky. This nonuniformity was found in 1992 by the Cosmic Background Explorer (COBE) satellite, a finding later confirmed by several other experiments. Inflation also predicts that the universe should be nearly flat. Flatness of the universe can be experimentally verified because the density of a flat universe is related in a simple way to the speed of its expansion. So far observational data are consistent with this prediction. A few years ago it seemed that if someone were to show that the universe is open rather than flat, then inflationary theory would fall apart. Recently, however, several models of an open inflationary universe have been found. The only consistent description of a large homogeneous open universe that we currently know is based on inflationary theory. Thus, even if the universe is open, inflation is still the best theory to describe it. One may argue that the only way to disprove the theory of inflation is to propose a better theory.

One should remember that inflationary models are based on the theory of elementary particles, and this theory is not completely established. Some versions (most notably, superstring theory) do not automatically lead to inflation. Pulling inflation out of the superstring model may require radically new ideas.

Here we come to the most interesting part of our story, to the theory of an eternally existing, self-reproducing inflationary universe. This theory is rather general, but it looks especially promising and leads to the most dramatic consequences in the context of the chaotic inflation scenario.

As I already mentioned, one can visualize quantum fluctuations of the scalar field in an inflationary universe as waves. They first moved in all possible directions and then froze on top of one another. Each frozen wave slightly increased the scalar field in some parts of the universe and decreased it in others. Now consider those places of the universe where these newly frozen waves persistently increased the scalar field. Such regions are extremely rare, but still they do exist. And they can be extremely important. Those rare domains of the universe where the field jumps high enough begin exponentially expanding with ever-increasing speed. The higher the scalar field jumps, the faster the universe expands. Very soon those rare domains will acquire a much greater volume than other domains.

From this theory it follows that if the universe contains at least one inflationary domain of a sufficiently large size, it begins unceasingly producing new inflationary domains. Inflation in each particular point may end quickly, but many other places will continue to expand. The total volume of all these domains will grow without end. In essence, one inflationary universe sprouts other inflationary bubbles, which in turn produce other

Inside each bubble, a separate universe: An inflationary cosmos engenders many universes, each contained within its own bubble. Each local universe operates on its own set of physical laws, which may or may not be similar to the laws in our own universe.

inflationary bubbles. This process, which I have called eternal inflation, keeps going as a chain reaction, producing a fractallike pattern of universes. In this scenario the universe as a whole is immortal. Each particular part of the universe may stem from a singularity somewhere in the past, and it may end up in a singularity somewhere in the future. There is, however, no end for the evolution of the entire universe.

The situation with the very beginning is less certain. There is a chance that all parts of the universe were created simultaneously in an initial big bang singularity. The necessity of this assumption, however, is no longer obvious. Furthermore, the total number of inflationary bubbles on our "cosmic tree" grows exponentially in time. Therefore, most bubbles (including our own part of the universe) grow indefinitely far away from the trunk of this tree. Although this scenario makes the existence of the initial big bang almost irrelevant, for all practical purposes, one can consider the moment of formation of each inflationary bubble as a new "big bang."

A New Cosmology

Could matters become even more curious? The answer is yes. Until now, we have considered the simplest inflationary model with only one scalar field, which has only one minimum of its potential energy. Meanwhile realistic models of elementary particles propound many kinds of scalar fields. For example, in the unified theories of weak, strong and electromagnetic interactions, at least two other scalar fields exist. The potential energy of these scalar fields may have several different minima. This condition means that the same theory may have different "vacuum states," corresponding to different types of symmetry breaking between fundamental interactions and, as a result, to different laws of low-energy physics. (Interactions of particles at extremely large energies do not depend on symmetry breaking.)

Such complexities in the scalar field mean that after inflation the universe may become divided into exponentially large domains that have different laws of low-energy physics. Note that this division occurs even if the entire universe originally began in the same state, corresponding to one particular minimum of potential energy. Indeed, large quantum fluctuations can cause scalar fields to jump out of their minima. That is, they jiggle some of the balls out of their bowls and into other ones. Each bowl corresponds to alternative laws of particle interactions. In some inflationary models, quantum fluctuations are so strong that even the number of dimensions of space and time can change.

If this model is correct, then physics alone cannot provide a complete explanation for all properties of our allotment of the universe. Does this mean that understanding all the properties of our region of the universe will require, besides a knowledge of physics, a deep investigation of our own nature, perhaps even including the nature of our consciousness? This conclusion would certainly be one of the most unexpected that one could draw from the recent developments in inflationary cosmology.

The evolution of inflationary theory has given rise to a completely new cosmological paradigm, which differs considerably from the old big bang theory and even from the first versions of the inflationary scenario. In it the universe appears to be both chaotic and homogeneous, expanding and stationary. Our cosmic home grows, fluctuates and eternally reproduces itself in all possible forms, as if adjusting itself for all possible types of life.

Some parts of the new theory, we hope, will stay with us for years to come. Many others will have to be considerably modified to fit with new observational data and with the ever-changing theory of elementary particles.

The cosmic microwave background radiation that permeates the cosmos is one of our most valuable clues in determining what happened in the early stages of the universe's development. Trapped in the cosmic plasma of elementary particles for some 500,000 years after the big bang and then suddenly released, this radiation carries with it snapshots of the environment that gave birth to the universe. But it is not sufficient. The holy grail for cosmologists would be the discovery of gravity waves, because this as-yet-undetected form of energy would have traveled unimpeded through the early soup of subatomic particles, stretching and squeezing not only the cosmic microwave background radiation but the fabric of space-time itself. If found, gravitational waves would give cosmologists a glimpse of the big bang from its earliest moments.

Echoes from the Big Bang

Robert R. Caldwell and Marc Kamionkowski

T heorists have long speculated on the origin of the cosmos, but until recently, they had no way to probe the universe's earliest moments to test their hypotheses. In recent years, however, researchers have identified a method for observing the universe as it was in the very first fraction of a second after the big bang. This method involves looking for traces of gravitational waves in the cosmic microwave background (CMB), the cooled radiation that has permeated the universe for nearly 15 billion years.

In particular, researchers are hoping to find direct evidence of the epoch of inflation. The strongest evidence—the "smoking gun"—would be the observation of inflationary gravitational waves. Gravitational waves are moving disturbances of a gravitational field. Like light or radio waves, gravitational waves can carry information and energy from the sources that produce them.

The plasma that filled the universe during its first 500,000 years was opaque to electromagnetic radiation, because any emitted photons were immediately scattered in the soup of subatomic particles. Therefore, astronomers cannot observe any electromagnetic signals dating from before the CMB. In contrast, gravitational waves could propagate through the plasma. What is more, the theory of inflation predicts that the explosive expansion of the universe 10^{-38} second after the big bang should have produced gravitational waves. If the theory is correct, these waves would have echoed across the early universe and, 500,000 years later, left subtle ripples in the CMB that can be observed today.

Waves from Inflation

To understand how inflation could have produced gravitational waves, let's examine a fascinating consequence of quantum mechanics: empty space is not so empty. Virtual pairs of particles are spontaneously created and destroyed all the time. The Heisenberg uncertainty principle declares that a pair of particles may pop into existence for a time before they annihilate each other. You need not worry, though, about virtual apples or bananas popping out of empty space, because the formula applies only to elementary particles and not to complicated arrangements of atoms.

One of the elementary particles affected by this process is the graviton, the quantum particle of gravitational waves. Pairs of virtual gravitons are constantly popping in and out of existence. During inflation, however, the virtual gravitons would have been pulled apart much faster than they could have disappeared back into the vacuum. In essence, the virtual particles would have become real particles. Furthermore, the fantastically rapid expansion of the universe would have stretched the graviton wavelengths from microscopic to macroscopic lengths. In this way, inflation would have pumped energy into the production of gravitons, generating a spectrum of gravitational waves that reflected the conditions in the universe in those first moments after the big bang. If inflationary gravitational waves do indeed exist, they would be the oldest relic in the universe, created 500,000 years before the CMB was emitted.

Whereas the microwave radiation in the CMB is largely confined to wave-lengths between one and five millimeters (with a peak intensity at two millimeters), the wavelengths of the inflationary gravitational waves would span a much broader range: one centimeter to 10^{-23} kilometers, which is the size of the present-day observable universe. The theory of inflation stipulates that the gravitational waves with the longest wavelengths would be the most intense and that their strength would depend on the rate at which the universe expanded during the inflationary epoch. This rate is proportional to the energy scale of inflation, which was determined by the temperature of the universe when inflation began. And because the universe was hotter at earlier times, the strength of the gravitational waves ultimately depends on the time at which inflation started.

Unfortunately, cosmologists cannot pinpoint this time, because they do not know in detail what caused inflation. Some physicists have theorized that inflation started when three of the fundamental interactions—the strong, weak, and electromagnetic forces—became dissociated soon after the universe's creation. According to this theory, the three forces were one and the same at the very beginning but became distinct 10^{-38} second after the big bang, and this event somehow triggered the sudden expansion of the cosmos. On the other hand, if inflation were triggered by another physical phenomenon occurring at a later time, the gravitational waves would be weaker.

Once produced during the first fraction of a second after the big bang, the inflationary gravitational waves would propagate forever, so they should still be running across the universe. But how can cosmologists observe them? First consider how an ordinary stereo receiver detects a radio signal. The radio waves consist of oscillating electrical and magnetic fields, which cause the electrons in the receiver's antenna to move back and forth. The motions of these electrons produce an electric current that the receiver records.

Researchers are now building sophisticated gravitational wave detectors, which will use lasers to track the tiny motions of suspended masses. The distance between the test masses determines the band of wavelengths that the devices can monitor. The largest of the ground-based detectors, which has a separation of four kilometers between the masses, will be able to measure the oscillations caused by gravitational waves with wave lengths from 30 to 30,000 kilometers; a planned space-based observatory may be able to detect wavelengths about 1,000 times longer. The gravitational waves generated by neutron star mergers and black hole collisions have wave lengths in this range, so they can be detected by the new instruments. But the inflationary gravitational waves in this range are much too weak to produce measurable oscillations in the detectors.

The strongest inflationary gravitational waves are those with the longest wavelengths, comparable to the diameter of the observable universe. To detect these waves, researchers need to observe a set of freely floating test masses separated by similarly large distances. Serendipitously, nature has provided just such an arrangement: the primordial plasma that emitted the CMB radiation. During the 500,000 years between the epoch of inflation and the emission of the CMB, the ultralong wavelength gravitational waves echoed across the early universe, alternately stretching and squeezing the plasma. Researchers can observe these oscillatory motions today by looking for slight Doppler shifts in the CMB.

If, at the time when the CMB was emitted, a gravitational wave was stretching a region of plasma toward us—that is, toward the part of the universe that would eventually become our galaxy—the radiation from that region will appear bluer to observers because it has shifted to shorter wavelengths (and hence a higher temperature). Conversely, if a gravitational wave was squeezing a region of plasma away from us when the CMB was emitted, the radiation will appear redder because it has shifted to longer wavelengths (and a lower temperature). By surveying the blue and red spots in the CMB—which correspond to hotter and colder radiation temperatures—researchers could conceivably see the pattern of plasma motions induced by the inflationary gravitational waves. The universe itself becomes a gravitational-wave detector.

The Particulars of Polarization

The task is not so simple, however. As we noted at the beginning of this article, mass inhomogeneities in the early universe also produced temperature variations in the CMB. (For example, the gravitational field of the denser regions of plasma would have redshifted the photons emitted from those regions, producing some of the temperature differences observed by COBE.) If cosmologists look at the radiation temperature alone, they cannot tell what fraction (if any) of the variations should be attributed to gravitational waves. Even so, scientists at least know that gravitational waves could not have produced any more than the one-in-100,000 temperature differences observed by COBE and the other CMB radiation detectors.

But how can cosmologists go further? How can they get around the uncertainty over the origin of the temperature fluctuations? The answer lies with the polarization of the CMB. When light strikes a surface in such a way that the light scatters at nearly a right angle from the original beam, it becomes linearly polarized—that is, the waves become oriented in a particular direction. This is the effect that polarized sunglasses exploit: because the sunlight that scatters off the ground is typically polarized in a horizontal direction, the filters in the glasses reduce the glare by blocking light waves with this ori-

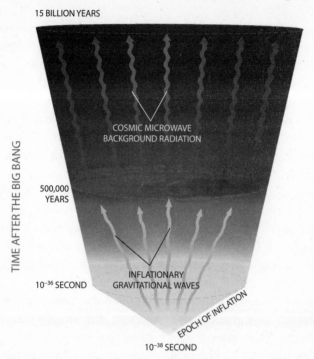

15 BILLION YEARS

TIME AFTER THE BIG BANG

COSMIC MICROWAVE
BACKGROUND RADIATION

500,000
YEARS

10^{-36} SECOND

INFLATIONARY
GRAVITATIONAL WAVES

EPOCH OF INFLATION

10^{-38} SECOND

Just 10^{-38} seconds after the big bang, gravity waves permeated the universe, followed half a million years later by cosmic microwave background radiation. CMB may subtly reflect the patterns of the elusive gravity wave.

entation. The CMB is polarized as well. Just before the early universe became transparent to radiation, the CMB photons scattered off the electrons in the plasma for the last time. Some of these photons struck the particles at large angles, which polarized the radiation.

The key to detecting the inflationary gravitational waves is the fact that the plasma motions caused by the waves produced a different pattern of polarization than the mass inhomogeneities did. The idea is relatively simple. The linear polarization of the CMB can be depicted with small line segments that show the orientation angle of the polarization in each region of the sky. These line segments are sometimes arranged in rings or in radial patterns. The segments can also appear in rotating swirls that are either right- or left-handed— that is, they seem to be turning clockwise or counterclockwise.

The "handedness" of these latter patterns is the clue to their origin. The mass inhomogeneities in the primordial plasma could not have produced such polarization patterns, because the dense and rarefied regions of plasma had no right- or left-handed orientation. In contrast, gravitational waves do have a handedness: they propagate with either a right- or left-handed screw motion. The polarization pattern produced by gravitational waves will look like a random superposition of many rotating swirls of various sizes. Researchers describe these patterns as having a curl, whereas the ringlike and radial patterns produced by mass inhomogeneities have no curl.

Not even the most keen-eyed observer can look at a polarization diagram and tell by eye whether it contains any patterns with curls. But an extension of Fourier analysis—a mathematical technique that can break up an image into a series of waveforms—can be used to divide a polarization pattern into its constituent curl and curl-free patterns. Thus, if cosmologists can measure the CMB polarization and determine what fraction came from curl patterns, they can calculate the amplitude of the ultra-long wavelength inflationary gravitational waves. Because the amplitude of the waves was determined by the energy of inflation, researchers will get a direct measurement of that energy scale. This finding, in turn, will help answer the question of whether inflation was triggered by the unification of fundamental forces.

What are the prospects for the detection of these curl patterns? NASA's MAP spacecraft and several ground-based and balloon-borne experiments are poised to measure the polarization of the CMB for the very first time, but these instruments will probably not be sensitive enough to detect the curl component produced by inflationary gravitational waves. Subsequent

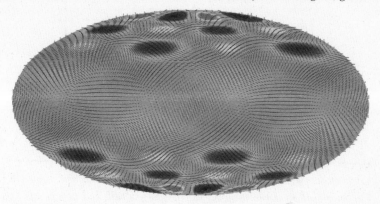

With gravity waves bending and stretching the fabric of the cosmic microwave background radiation, temperature variations formed, indicated by the dark areas in this simulation. The cross-hatch lines represent the direction of polarization in different regions of the sky.

experiments may have a better chance, though. If inflation was indeed caused by the unification of forces, its gravitational-wave signal might be strong enough to be detected by the Planck spacecraft, although an even more sensitive next-generation spacecraft might be needed. But if inflation was triggered by other physical phenomena occurring at later times and lower energies, the signal from the gravitational waves would be far too weak to be detected in the foreseeable future.

Because cosmologists are not certain about the origin of inflation, they cannot definitively predict the strength of the polarization signal produced by inflationary gravitational waves. But if there is even a small chance that the signal is detectable, then it is worth pursuing. Its detection would not only provide incontrovertible evidence of inflation but also give us the extraordinary opportunity to look back at the very earliest times, just 10^{-38} second after the big bang. We could then contemplate addressing one of the most compelling questions of the ages: Where did the universe come from?

A famous parable from ancient times tells of three blind men led to an elephant. One touches the elephant's tail and proclaims he is holding a rope. Another, grabbing the elephant's trunk, announces he is in contact with a snake. The third, stroking the elephant's side, believes he is touching a wall. In the same spirit, perhaps the partial picture we've gleaned from our investigation of the universe is an illusion based on limits imposed by our technologies and our preconceptions.

Using Einstein's general theory of relativity as a springboard, along with puzzling measurements suggesting far less matter in the cosmos than predicted by the inflation theory, the following essay explores a universe of a different shape. Here we delve into the possibility of the cosmos being curved like a bubble and the implications of a bubble-shaped universe.

some fits and starts, the
ry seemed to best accommoda
growing observations. Tiny
tuations found throughout t
ic microwave background ech
erse, s hints one of t
y hallmark proofs of the bi
rs soon followed. Most mode
ological thought frames its
big bang theory when its

Inflation in a Low-Density Universe

Martin A. Bucher and David N. Spergel

osmology has a reputation as a difficult science, but in many ways explaining the whole universe is easier than understanding a single-celled animal. On the largest cosmic scales, where stars, galaxies and even galaxy clusters are mere flecks, matter is spread out evenly. And it is governed by only one force, gravity. These two basic observations— large-scale uniformity and the dominance of gravity—are the basis of the big bang theory, according to which our universe has been expanding for the past 12 billion years or so. Despite its simple underpinnings, the theory is remarkably successful in explaining the velocity of galaxies away from one another, the relative amounts of light elements, the dim microwave glow in the sky and the general evolution of astronomical structures. The unfolding of the cosmos, it seems, is almost completely insensitive to the details of its contents. Unfortunately for biologists, the same principle does not apply to even the simplest organism.

Yet there are paradoxes inherent in the big bang theory. Two decades ago cosmologists resolved these troubling inconsistencies by incorporating ideas from particle physics—giving rise to the theory of "inflation." But now this elaboration is itself facing a crisis, brought on by recent observations that contradict its prediction for the average density of matter in the cosmos. Cosmologists are realizing that the universe may not be quite so simple as they had thought. Either they must posit the existence of an exotic form of matter or energy, or they must add a layer of complexity to the theory of inflation. In this article we will focus on the second option.

Strictly speaking, the big bang theory does not describe the birth of the universe, but rather its growth and maturation. According to the theory, the infant universe was an extremely hot, dense cauldron of radiation. A part of it, a chunk smaller than a turnip, eventually enlarged into the universe observable today

A natural consequence of the expansion of a uniform universe is Hubble's Law, whereby galaxies are moving away from the Earth (or from any other point in the universe) at speeds proportional to their distance. Not all objects in the universe obey this law, because mutual gravitational attraction fights against the swelling of space. For example, the sun and the Earth are not moving apart. But it holds on the largest scales. In the simplest version of the big bang, the expansion has always proceeded at much the same rate.

In the Beginning, Paradox

Despite its successes, the standard big bang theory cannot answer several profound questions. First, why is the universe so uniform? Two regions on opposite sides of the sky look broadly the same, yet they are separated by more than 24 billion light-years. Light has been traveling for only about 12 billion years, so the regions have yet to see each other. There has never been enough time for matter, heat or light to flow between them and homogenize their density and temperature.

Somehow the uniformity of the universe must have predated the expansion, but the theory does not explain how.

Conversely, why did the early universe have any density variations at all? Fortunately, it did: without these tiny undulations, the universe today would still be of uniform density—a few atoms per cubic meter—and neither the Milky Way nor the earth would exist.

Finally, why is the rate of cosmic expansion just enough to counteract the collective gravity of all the matter in the universe? Any significant deviation from perfect balance would have magnified itself over time. If the expansion rate had been too large, the universe today would seem nearly devoid of matter. If gravity had been too strong, the universe would have already collapsed in a big crunch, and you would not be reading this article.

Cosmologists express this question in terms of the variable omega, Ω, the ratio of gravitational energy to kinetic energy (that is, the energy contained in the motion of matter as space expands). The variable is proportional to the density of matter in the universe—a higher density means stronger gravity, hence a larger Ω. If Ω equals one, its value never changes; otherwise it rapidly decreases or increases in a self-reinforcing process, as either kinetic or gravitational energy comes to dominate. After billions of years, Ω should effectively be either zero or infinity. Because the current density of the universe is (thankfully) neither zero nor infinity, the original value of Ω must have been exactly one or extraordinarily close to it (within one part in 10^{18}). Why? The big bang theory offers no explanation apart from dumb luck.

These shortcomings do not invalidate the theory—which neatly explains billions of years of cosmic history—but they do indicate that it is incomplete. To fill in the gap, in the early 1980s cosmologists Alan H. Guth, Katsuhiko Sato, Andrei D. Linde, Andreas Albrecht and Paul J. Steinhardt developed the theory of inflation.

How Did the Universe Begin?

The laws of physics generally describe how a physical system develops from some initial state. But any theory that explains how the universe began must involve a radically different kind of law, one that explains the initial state itself. If normal laws are road maps telling you how to get from A to B, the new laws must justify why you started at A to begin with. Many creative possibilities have been proposed.

In 1983 James B. Hartle of the University of California at Santa Barbara and Stephen W. Hawking of the University of Cambridge applied quantum mechanics to the universe as a whole, producing a cosmic wave function analogous to the wave function for atoms and elementary particles. The wave function determines the initial conditions of the universe. According to this approach, the usual distinction between future and past breaks down in the very early universe; the time direction takes on the properties of a spatial direction. Just as there is no edge to space, there is no identifiable beginning to time. In an alternative hypothesis, Alexander Vilenkin of Tufts University proposed a "tunneling" wave function determined by the relative probabilities for a universe of zero size to become a universe of finite size of its own accord.

Last year Hawking and Neil G. Turok, also at Cambridge, suggested the spontaneous creation of an open inflationary bubble from nothingness. This new version of open inflation bypasses the need for false-vacuum decay, but Vilenkin and Andrei D. Linde of Stanford University have challenged the assumptions in the calculation.

Linde has tried to skirt the problem of initial conditions by speculating that inflation is a process without beginning [see "The Self-Reproducing Inflationary Universe," page 32]. In the classical picture, inflation comes to an end as the inflation field rolls down its potential. But because of quantum fluctuations, the field can jump up the potential as well as down. Thus, there are always regions of the universe—in fact, constituting a majority of its volume—that are inflating. They surround pockets of space where inflation has ended and a stable universe has unfolded. Each pocket has a different set of physical constants; we live in the one whose constants are suited for our existence. The rest of the universe carries on inflating and always has. But Vilenkin and Arvind Borde, also at Tufts, have argued that even this extension of inflation does not describe the origin of the universe completely. Although inflation can be eternal in the forward time direction, it requires an ultimate beginning.

J. Richard Gott II and Li-Xin Li of Princeton University recently proposed that the universe is trapped in a cyclic state, rather like a time traveler who goes back in time and becomes her own mother. Such a person has no family tree; no explanation of her provenance is possible. In Gott and Li's hypothesis, our bubble broke off from the cyclic proto-universe; it is no longer cyclic but instead is always expanding and cooling.

Unfortunately, it may be very difficult (though perhaps not impossible) for astronomers to test any of these ideas. Inflation erases almost all observational signatures of what preceded it. Many physicists suspect that a fuller explanation of the preinflationary universe—and of the origin of the physical laws themselves—will have to await a truly fundamental theory of physics, perhaps string theory.

—M.A.B. and D.N.S.

The price paid for resolving the paradoxes is to make big bang theory more complicated. The inflationary theory postulates that the baby universe went through a period of very rapid expansion (hence the name). Unlike conventional big bang expansion, which decelerates over time, the inflationary expansion accelerated. It pushed any two independent objects apart at an ever-increasing clip—eventually faster than light. This motion did not violate relativity, which prohibits bodies of finite mass from moving through space faster than light. The objects, in fact, stood still relative to the space around them. It was space itself that came to expand faster than light.

Such rapid expansion early on explains the uniformity of the universe seen today. All parts of the visible universe were once so close together that they were able to attain a common density and temperature. During inflation, different parts of this uniform universe fell out of touch; only later, after inflation ended, did light have time to catch up with the slower, big bang expansion. If there is any non-uniformity in the broader universe, it has yet to come into view.

Field Work

To bring about the rapid expansion, inflationary theory adds a new element to cosmology, drawn from particle physics: the "inflaton" field. In modern physics, elementary particles, such as protons and electrons, are represented by quantum fields, which resemble the familiar electric, magnetic and gravitational fields. A field is simply a function of space and time whose oscillations are interpreted as particles. Fields are responsible for the transmission of forces.

The inflaton field imparts an "antigravity" force that stretches space. Associated with a given value of the inflaton field is a potential energy. Much like a ball rolling down a hill, the inflaton field tries to roll toward the bottom of its potential. But the expansion of the universe introduces what may be

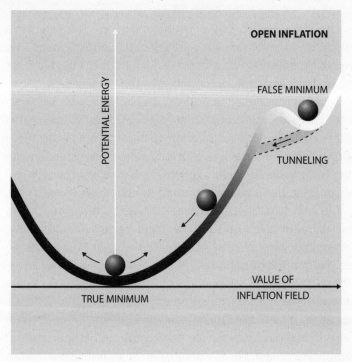

In the open inflation model, a universe grows when its expansion comes to a false minimum. During this resting phase, quantum physics explains that a tunneling action can take place, dropping the energy to a new slope where the cycle of creation begins anew.

described as a cosmological friction, impeding the descent. As long as the friction dominates, the inflaton field is almost stuck in place. Its value is nearly constant, so the anti-gravity force gains in strength relative to gravity—causing the distance between once nearby objects to increase at ever faster rates. Eventually the field weakens and converts its remaining energy into radiation. Afterward the expansion of the universe continues as in the standard big bang.

Cosmologists visualize this process in terms of the shape of the universe. According to Einstein's general theory of relativity, gravity is a geometric effect: matter and energy warp the fabric of space and time, distorting the paths that objects follow. The overall expansion of the universe, which itself is a kind of bending of space and time, is controlled by the value of Ω.

If Ω is greater than one, the universe has a positive curvature, like the surface of an orange but in three spatial dimensions (the spherical, or "closed," geometry). If Ω is less than one, the universe has a negative curvature, like a potato chip (the hyperbolic, or "open," geometry). If it equals one, the universe is flat, like a pancake (the usual Euclidean geometry).

Inflation flattens the observable universe. Whatever the initial shape of the universe, the rapid expansion bloats it to colossal size and pushes most of it out of sight. The small visible fraction might seem flat, just as a small part of the earth's surface seems flat. Inflation thus pushes the observed value of Ω toward one. At the same time, any initial irregularities in the density of matter and radiation get evened out.

So in standard inflationary theory, cosmic flatness and uniformity are linked. For the universe to be as homogeneous as it is, the theory says the universe should be very, very flat, with Ω equal to one within one part in 100,000. Any deviation from exact flatness should be utterly impossible for astronomers to detect. Thus, for most of the past two decades observational flatness has been viewed as a firm prediction of the theory. And that is the problem. A wide variety of astronomical observations, involving galaxy clusters and distant supernovae, now suggest that gravity is too weak to combat the expansion. If so, the density of matter must be less than predicted—with Ω equal to about 0.3. That is, the universe might be curved and open. There are three ways to interpret this result. The first is that inflationary theory is completely wrong. But if cosmologists abandon inflation, the formidable paradoxes so nicely resolved by the theory would reappear, and a new theory would be required. No such alternative is known.

A second interpretation takes heart from the accelerating

expansion inferred from the observations of distant super-novae Such expansion hints at additional energy in the form of a "cosmological constant." This extra energy would act as a weird kind of matter, bending space much as ordinary matter does. The combined effect would be to flatten space, in which case the inflationary theory has nothing to worry about [see "Cosmological Antigravity," page 88]. But the inference of the cosmological constant is plagued by uncertainties about dust and the nature of the stars that undergo supernova explosions. So cosmologists are keeping their options open (so to speak).

Bubble Universe

A third path is to take the observations at face value and ask whether a flat universe really is an inevitable consequence of inflation. This approach involves yet another extension of the theory to still earlier times, with some new complexity. The route was first mapped in the early 1980s by Sidney R. Cole-man and Frank de Luccia of Harvard University and J. Richard Gott II of Princeton University. Ignored for over a decade, the ideas were recently developed by one of us (Bucher), along with Neil G. Turok, now at the University of Cambridge, and Alfred S. Goldhaber of the State University of New York at Stony Brook, and by Misao Sasaki and Takahiro Tanaka, now at Osaka University, and Kazuhiro Yamamoto of Kyoto University. Linde and his collaborators have also proposed some concrete models and extensions of these ideas.

If the inflaton field had a different potential-energy function, inflation would have bent space in a precise and predictable way—leaving the universe slightly curved rather than exactly flat. In particular, suppose the potential-energy function had two valleys—a false (local) minimum as well as a true (global) mini-mum. As the inflaton field rolled down, the universe expanded and became uniform. But then the field got stuck in the false

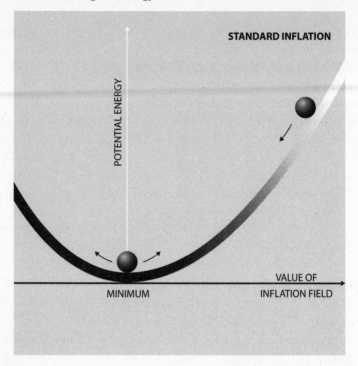

STANDARD INFLATION

POTENTIAL ENERGY

MINIMUM

VALUE OF
INFLATION FIELD

The origin of the force that caused space to expand behaved like a ball rolling down a hill. Once in this energy valley, the force filled space with matter and radiation, triggering the big bang.

minimum. Physicists call this state the "false vacuum," and any matter and radiation in the cosmos were almost entirely replaced by the energy of the inflaton field. The fluctuations inherent in quantum mechanics caused the inflaton field to jitter and ultimately enabled it to escape from the false minimum—just as shaking a pinball machine can free a trapped ball.

The escape, called false-vacuum decay, did not occur everywhere at the same time. Rather it first took place at some random location and then spread. The process was analogous to bringing water to a boil. Water heated to its boiling point does not instantaneously turn into steam everywhere. First, because of the random motion of atoms, scattered bubbles nucleate throughout the liquid—rather like the burbling of a pot of soup. Bubbles smaller than a certain minimum size collapse because of surface tension. But in larger bubbles, the energy difference between the steam and the super-heated water overcomes surface tension; these bubbles expand at the speed of sound in water.

In false-vacuum decay, quantum fluctuations played the role of the random atomic motion, causing bubbles of true vacuum to nucleate. Surface tension destroyed most of the bubbles, but a few managed to grow so large that quantum effects became unimportant. With nothing to oppose them, their radius continued to increase at the speed of light. As the outside wall of a bubble passed through a point in space, the inflaton field at that point was jolted out of the false minimum and resumed its downward descent. Thereafter the space inside the bubble inflated much as in standard inflationary theory. The interior of this bubble corresponds to our universe. The moment that the inflaton field broke out of its false minimum corresponds to the big bang in older theories. For points at different distances from the center of nucleation, the big bang occurred at different times. This disparity seems strange, to say the least. But careful examination of the inflaton field reveals what went on. The inflaton acted as a chronometer: its value at a given point represented the time elapsed since the big bang occurred at that point. Because of the time lag in the commencement of the big bang, the value of the inflaton was not the same everywhere; it was highest at the wall of the bubble and fell steadily toward the center.

The value of the inflaton is no mere abstraction. It determined the basic properties of the universe inside the bubble—namely, its average density and the temperature of the cosmic background radiation (today 2.7 degrees C above absolute zero). Along a hyperbolic surface, the density, temperature and elapsed time were constant. These surfaces are what observers inside the bubble perceive as constant "time." It is not the same as time experienced outside the bubble.

How is it possible for something so fundamental as time to be different on the inside and on the outside? Based on the understanding of space and time before Einstein's theories of relativity, such a feat would indeed have seemed impossible. But in relativity, the distinction between space and time blurs. What any observer calls "space" and "time" is largely a matter of convenience. Loosely speaking, time represents the direction in which things change, and change inside the bubble is driven by the inflaton.

Bounded in a Nutshell

According to relativity, the universe has four dimensions—three for space, one for time. Once the direction of time is determined, the three remaining directions must be spatial; they are the directions in which time is constant. Therefore, a bubble universe seems hyperbolic from the inside. For us, to travel out in space is, in effect, to move along a hyperbola. To look backward in time is to look toward the wall of the bubble. In principle, we could look outside the bubble and before the big bang, but in practice, the dense, opaque early universe blocks the view.

This melding of space and time allows an entire hyperbolic universe (whose volume is infinite) to fit inside an expanding bubble (whose volume, though increasing without limit, is always finite). The space inside the bubble is actually a blend of both space and time as perceived outside the bubble. Because external time is infinite, so is internal space. The seemingly bizarre concept of bubble universes frees inflationary theory from its insistence that Ω equal one. Although the formation of the bubble created hyperbolas, it said nothing about their precise scale. The scale is instead determined by the details of the inflaton potential, and it varies over time in accordance with the value of Ω. Initially, Ω inside the bubble equals zero. During inflation, its value increases, approaching one. Thus, hyperbolas start off with an abrupt bend and gradually flatten out. The inflaton potential sets the rate and duration of flattening. Eventually inflation in the bubble comes to an end, at which point Ω is poised extremely near but very slightly below one. Then Ω starts to decrease. If the duration of inflation inside the bubble is just right (to within a few percent), the current value of Ω will match the observed value.

At first glance the process may seem baroque, but the main conclusion is simple: the uniformity and geometry of the universe need not be linked. Instead they could result from different stages of inflation: uniformity, from inflation before the nucleation of the bubble; geometry, from inflation within the bubble. Because the two properties are not intertwined, the need for uniformity does not determine the duration of inflation, which lasts just long enough to give the hyperbolas the desired degree of flatness.

In fact, this formulation is a straightforward extension of the big bang theory. The standard view of inflation describes what happened just before the conventional big bang expansion. The new conception, known as open inflationary theory, adds another stage preceding standard inflation. Another theory describing even earlier times will be needed to explain the original creation of the universe [See "Quantum Cosmology and the Creation of the Universe," page 100].

Life in a bubble universe has a number of interesting consequences (not to mention possibilities for science fiction plots). For instance, an alien observer could safely pass from the outside to the inside of the bubble. But once inside, the observer (like us) could never leave, for doing so would require traveling faster than light. Another implication is that our universe is only one of an infinity of bubbles immersed in a vast, frothy sea of eternally expanding false vacuum. What if two bubbles col-

The Geometry of the Universe

If the universe had an "outside" and people could view it from that perspective, cosmology would be much easier. Lacking these gifts, astronomers must infer the basic shape of the universe from its geometric properties. Everyday experience indicates that space is Euclidean, or "flat," on small scales. Parallel lines never meet, triangles span 180 degrees, the circumference of a circle is $2\pi r$, and so on. But it would be wrong to assume that the universe is Euclidean on large scales, just as it would be wrong to conclude that the earth is flat just because a small patch of it looks flat.

There are two other possible three-dimensional geometries consistent with the observations of cosmic homogeneity (the equivalence of all points in space) and isotropy (the equivalence of all directions). They are the spherical, or "closed," geometry and the hyperbolic, or "open," geometry. Both are characterized by a curvature length analogous to the Earth's radius. If the curvature is positive, the geometry is spherical; if negative, hyperbolic. For distances much smaller than this length, all geometries look Euclidean.

In a spherical universe, as on the Earth's surface, parallel lines eventually meet, triangles can span up to 540 degrees, and the circumference of a circle is smaller than $2\pi r$. Because the space curves back on itself, the spherical universe is finite. In a hyperbolic universe, parallel lines diverge, triangles have less than 180 degrees, and the circumference of a circle is larger than $2\pi r$. Such a universe, like Euclidean space, is infinite in size. (There are ways to make hyperbolic and flat universes finite, but they do not affect the conclusions of inflationary theory.)

These three geometries have quite different effects on perspective, which distort the appearance of features in the cosmic microwave background radiation. The largest ripples in the background have the same absolute size regardless of the specific process of inflation. If the universe is flat, the largest undulations would appear to be about one degree across. But if the universe is hyperbolic, the same features should appear to be only half that size, simply because of geometric distortion of light rays.

Preliminary observations hint that the ripples are indeed one degree across. If confirmed, these results imply that the open inflationary theory is wrong. But tentative findings are often proved wrong, so astronomers await upcoming satellite observations for a definitive answer.

—M.A.B. and D.N.S.

lided? Their meeting would unleash an explosion of cosmic proportions, destroying everything inside the bubbles near the point of impact. Fortunately, because the nucleation of bubbles is an extremely rare process, such cataclysms are improbable. Even if one occurred, a substantial portion of the bubbles would not be affected. To observers inside the bubbles but at a safe distance, the event would look like a broiling hot region in the sky.

Corroborating Evidence

How does one test this theory? To explain why the universe is uniform is certainly a good thing. But validating a theory requires that some quantitative predictions be compared with observations. The specific effects of open inflation were calculated in 1994, with contributions by the two groups that refined the theory, as well as Bharat V. Ratra and P. James E. Peebles.

Both the old and the new concepts of inflation make definite forecasts based on quantum effects, which caused different points in space to undergo slightly different amounts of inflation. When inflation ended, some energy was left over in the inflaton field and became ordinary radiation—the fuel of the subsequent big bang expansion. Because the duration of inflation varied from place to place, so did the amount of residual energy and therefore the density of the radiation.

The cosmic background radiation provides a snapshot of these undulations. In open inflation, it is affected not only by fluctuations that develop within the universe but also by ones that arise outside the bubble and propagate inside. Other ripples are set in motion by imperfections in the nucleation of the bubble. These patterns ought to be most notable on the largest scales. In effect, they allow us to look outside our bubble universe.

At the current level of precision, the observations cannot distinguish between the predictions of the two inflationary theories. The moment of truth will come with the deployment in 2001 of the Microwave Anisotropy Probe (MAP) by the National Aeronautics and Space Administration. A more advanced European counterpart, Planck, is due for launch in 2007. These satellites will perform observations similar to those of the Cosmic Microwave Background Explorer (COBE) satellite nearly a decade ago, but at much higher resolution. They should be able to pick out which theory—either the cosmological constant or open inflation—is correct. Or it could well turn out that neither fits, in which case researchers will have to start over and find some new ideas for what happened in the very early universe.

That the universe is expanding is a cosmological axiom. But what powers this outward expansion? Several candidates have been proposed, including vacuum energy, dark energy, and virtual particles. Yet each of these candidates comes with its own package of problems. One recently proposed and promising possibility for the energy source driving the universe to expand is called quintessence, *an energy field that mimics a constant energy vacuum, which resembles the inflaton fields proposed by the inflation theory. Yet unlike the inflaton field, quintessence may undergo diverse sorts of evolution.*

The Quintessential Universe

Jeremiah P. Ostriker
and Paul J. Steinhardt

I s it all over but the shouting? Is the cosmos understood aside from minor details? A few years ago it certainly seemed that way. After a century of vigorous debate, scientists had reached a broad consensus about the basic history of the universe. It all began with gas and radiation of unimaginably high temperature and density. For 15 billion years, it has been expanding and cooling. Galaxies and other complex structures have grown from microscopic seeds—quantum fluctuations—that were stretched to cosmic size by a brief period of "inflation." We had also learned that only a small fraction of matter is composed of the normal chemical elements of our everyday experience. The majority consists of so-called dark matter, primarily exotic elementary particles that do not interact with light. Plenty of mysteries remained, but at least we had sorted out the big picture.

Or so we thought. It turns out that we have been missing most of the story. Over the past five years, observations have convinced cosmologists that the chemical elements and the dark matter, combined, amount to less than half the contents of the universe. The bulk is a ubiquitous "dark energy" with a strange and remarkable feature: its gravity does not attract. It repels. Whereas gravity pulls the chemical elements and dark matter into stars and galaxies, it pushes the dark energy into a nearly uniform haze that permeates space. The universe is a battleground between the two tendencies, and repulsive gravity is winning. It is gradually overwhelming the attractive force of ordinary matter—causing the universe to accelerate to ever larger rates of expansion, perhaps leading to a new runaway inflationary phase and a totally different future for the universe than most cosmologists envisioned a decade ago.

Until recently, cosmologists have focused simply on proving the existence of dark energy. Having made a convincing case, they are now turning their attention to a deeper problem: Where does the energy come from? The best known possibility is that the energy is inherent in the fabric of space. Even if a volume of space were utterly empty—without a bit of matter and radiation—it would still contain this energy.

Later, when astronomers established that the universe is expanding, Einstein regretted his delicately tuned notion of a cosmological constant, calling it his greatest blunder. But perhaps his judgment was too hasty. If the cosmological constant had a slightly larger value than Einstein proposed, its repulsion would exceed the attraction of matter, and cosmic expansion would accelerate.

Many cosmologists, though, are now leaning toward a different idea, known as quintessence. The translation is "fifth element," an allusion to ancient Greek philosophy, which suggested that the universe is composed of earth, air, fire and water, plus an ephemeral substance that prevents the moon and planets from falling to the center of the celestial sphere. Three years ago Robert R. Caldwell, Rahul Dave and one of us (Steinhardt), all then at the University of Pennsylvania, reintroduced the term to refer to a dynamical quantum field, not unlike an electrical or magnetic field, that gravitationally repels.

The dynamism is what cosmologists find so appealing about quintessence. The biggest challenge for any theory of dark energy is to explain the inferred amount of the stuff—not so much that it would have interfered with the formation of stars and galaxies in the early universe but just enough that its effect can now be felt. Vacuum energy is completely inert, maintaining the same density for all time. Consequently, to explain the amount of dark energy today, the value of the cosmological constant would have to be fine tuned at the creation of the universe to have the proper value—which makes it sound rather like a fudge factor. In contrast, quintessence interacts with matter and evolves with time, so it might naturally adjust itself to reach the observed value today.

Two-Thirds of Reality

Distinguishing between these two options is critically important for physics. Particle physicists have depended on high-energy accelerators to discover new forms of energy and matter. Now the cosmos has revealed an unanticipated type of energy, too thinly spread and too weakly interacting for accelerators to probe. Whether the energy is inert or dynamical may be crucial to developing a fundamental theory of nature. Particle physicists are discovering that they must keep a close eye on developments in the heavens as well as in the accelerator laboratory.

The case for dark energy has been building brick by brick for nearly a decade. The first brick was a thorough census of all matter in galaxies and galaxy clusters using a variety of optical, x-ray and radio techniques. The unequivocal conclusion was that the total mass in chemical elements and dark matter accounts for only about one third of the quantity that most theorists expected—the so-called critical density.

Many cosmologists took this as a sign that the theorists were wrong. In that case, we would be living in an ever expanding universe where space is curved hyperbolically, like the horn on a trumpet [see "Inflation in a Low-Density Universe," page 55]. But this interpretation has been put to rest by measurements of hot and cold spots in the cosmic microwave background radiation, whose distribution has shown that space is flat and that the total energy density equals the critical density. Putting the two observations together, simple arithmetic dictates the necessity for an additional energy component to make up the missing two thirds of the energy density.

Whatever it is, the new component must be dark, neither absorbing nor emitting light, or else it would have been noticed long ago. In that way, it resembles dark matter. But the new component—called dark energy—differs from dark matter in one major respect: it must be gravitationally repulsive. Otherwise it would be pulled into galaxies and clusters, where it would affect the motion of visible matter. No such influence is seen. Moreover, gravitational repulsion resolves the "age crisis" that plagued cosmology in the 1990s. If one takes the current measurements of the expansion rate and assumes that the expansion has been decelerating, the age of the universe is less than 12 billion years.

Yet evidence suggests that some stars in our galaxy are 15 billion years old. By causing the expansion rate of the universe to accelerate, repulsion brings the inferred age of the cosmos into agreement with the observed age of celestial bodies

The potential flaw in the argument used to be that gravitational repulsion should cause the expansion to accelerate, which had not been observed. Then, in 1998, the last brick fell into place. Two independent groups used measurements of distant supernovae to detect a change in the expansion rate. Both groups concluded that the universe is accelerating and at just the pace predicted.

All these observations boil down to three numbers: the average density of matter (both ordinary and dark), the average density of dark energy and the curvature of space. Einstein's equations dictate that the three numbers add up to the critical density.

From Implosion to Explosion

Our everyday experience is with ordinary matter, which is gravitationally attractive, so it is difficult to envisage how dark energy could gravitationally repel. The key feature is that it's pressure negative. In Newton's law of gravity, pressure plays no role; the strength of gravity depends only on mass. In Einstein's law of gravity, however, the strength of gravity depends not just on mass but also on other forms of energy and on pressure. In this way, pressure has two effects: direct (caused by the action of the pressure on surrounding material) and indirect (caused by the gravitation that the pressure creates).

The sign of the gravitational force is determined by the algebraic combination of the total energy density plus three times the pressure. If the pressure is positive, as it is for radiation, ordinary matter and dark matter, then the combination is positive and gravitation is attractive. If the pressure is sufficiently negative, the combination is negative and gravitation is repulsive.

Vacuum energy meets this condition (provided its density is positive). This is a consequence of the law of conservation of energy, according to which energy cannot be destroyed. Mathematically the law can be rephrased to state that the rate of change of energy density is proportional to $w + 1$. For vacuum energy—whose density, by definition, never changes—this sum must be zero. In other words, w must equal precisely -1. So the pressure must be negative.

What does it mean to have negative pressure? Most hot gases have positive pressure; the kinetic energy of the atoms and radiation pushes outward on the container. Note that the direct effect of positive pressure—to push—is the opposite of its gravitational effect—to pull. But one can imagine an interaction among atoms that overcomes the kinetic energy and causes the gas to implode. The implosive gas has negative pressure. A balloon of this gas would collapse inward, because the outside pressure (zero or positive) would exceed the inside pressure (negative). Curiously, the direct effect of negative pressure—implosion—can be the opposite of its gravitational effect—repulsion.

Improbable Precision

The gravitational effect is tiny for a balloon. But now imagine filling all of space with the implosive gas. Then there is no bounding surface and no external pressure. The gas still has negative pressure, but it has nothing to push against, so it exerts no direct effect. It has only the gravitational effect—namely, repulsion. The repulsion stretches space, increasing its volume and, in turn, the amount of vacuum energy. The tendency to stretch is therefore self reinforcing. The universe expands at an accelerating pace. The growing vacuum energy comes at the expense of the gravitational field.

These concepts may sound strange, and even Einstein found them hard to swallow. He viewed the static universe, the original motivation for vacuum energy, as an unfortunate error that ought to be dismissed. But the cosmological constant, once introduced, would not fade away. Theorists soon realized that quantum fields possess a finite amount of vacuum energy, a manifestation of quantum fluctuations that conjure up pairs of "virtual" particles from scratch. An estimate of the total vacuum energy produced by all known fields predicts a huge amount—120 orders of magnitude more than the energy density in all other matter. That is, though it is hard to picture, the evanescent virtual particles should contribute a positive, constant energy density, which would imply negative pressure. But if this estimate were true, an acceleration of epic proportions would rip apart atoms, stars and galaxies. Clearly, the estimate is wrong. One of the major goals of unified theories of gravity has been to figure out why.

One proposal is that some heretofore undiscovered symmetry in fundamental physics results in a cancellation of large effects, zeroing out the vacuum energy. For example, quantum fluctuations of virtual pairs of particles contribute positive energy for particles with half integer spin (like quarks and electrons) but negative energy for particles with integer spin (like photons). In standard theories, the cancellation is inexact, leaving behind an unacceptably large energy density. But physicists have been exploring models with so-called supersymmetry, a relation between the two particle types that can lead to a precise cancellation. A serious flaw, though, is that supersymmetry would be valid only at very high energies. Theorists are working on a way of preserving the perfect cancellation even at lower energies.

Another thought is that the vacuum energy is not exactly nullified after all. Perhaps there is a cancellation mechanism that is slightly imperfect. Instead of making the cosmological constant exactly zero, the mechanism only cancels to 120 decimal places. Then the vacuum energy could constitute the missing two thirds of the universe. That seems bizarre, though. What mechanism could possibly work with such precision? Although the dark energy represents a huge amount of mass, it is spread so thinly that its energy is less than four electron volts per cubic millimeter—which, to a particle physicist, is unimaginably low. The weakest known force in nature involves an energy density 10^{50} times greater.

Extrapolating back in time, vacuum energy gets even more paradoxical. Today matter and dark energy have comparable average densities. But billions of years ago, when they came into being, our universe was the size of a grapefruit, so matter was 100 orders of magnitude denser. The cosmological constant, however, would have had the same value as it does now. In other words, for every 10^{100} parts matter, physical processes would have created one part vacuum energy—a degree of exactitude that may be reasonable in a mathematical idealization but that seems ludicrous to expect from the real world. This need for almost supernatural fine tuning is the principal motivation for considering alternatives to the cosmological constant.

Field Work

Fortunately, vacuum energy is not the only way to generate negative pressure. Another means is an energy source that, unlike vacuum energy, varies in space and time—a realm of possibilities that goes under the rubric of quintessence. For quintessence, w has no fixed value, but it must be less than $-\frac{1}{3}$ for gravity to be repulsive.

Quintessence may take many forms. The simplest models propose a quantum field whose energy is varying so slowly that it looks, at first glance, like a constant vacuum energy. The idea is borrowed from inflationary cosmology, in which a cosmic field known as the inflaton drives expansion in the very early universe using the same mechanism. The key difference is that quintessence is much weaker than the inflaton. This hypothesis was first explored a decade ago by Christof Wetterich of the University of Heidelberg and by Bharat Ratra, now at Kansas State University, and P. James E. Peebles of Princeton University.

In quantum theory, physical processes can be described in terms either of fields or particles. But because quintessence has such a low energy density and varies gradually, a particle of quintessence would be inconceivably lightweight and large—the size of a supercluster of galaxies. So the field description is rather more useful. Conceptually, a field is a continuous distribution of energy that assigns to each point in space a numerical value known as the field strength. The energy embodied by the field has a kinetic component, which depends on the time variation of the field strength, and a potential component, which depends only on the value of the field strength. As the field changes, the balance of kinetic and potential energy shifts.

In the case of vacuum energy, recall that the negative pressure was the direct result of the conservation of energy, which dictates that any variation in energy density is proportional to the sum of the energy density (a positive number) and the pressure. For vacuum energy, the change is zero, so the pressure must be negative. For quintessence, the change is gradual enough that the pressure must still be negative, though somewhat less so. This condition corresponds to having more potential energy than kinetic energy.

Because its pressure is less negative, quintessence does not accelerate the universe as strongly as vacuum energy does. Ultimately, this will be how observers decide between the two. If anything, quintessence is more consistent with the available data, but for now the distinction is not statistically significant. Another difference is that, unlike vacuum energy, the quintessence field may undergo all kinds of complex evolution. The value of w may be positive, then negative, then positive again. It may have different values in different places. Although the nonuniformity is thought to be small, it may be detectable by studying the cosmic microwave background radiation.

A further difference is that quintessence can be perturbed. Waves will propagate through it just as sound waves can pass through the air. In the jargon, quintessence is "soft." Einstein's cosmological constant is, in contrast, stiff—it cannot be pushed around. This raises an interesting issue. Every known form of energy is soft to some degree. Perhaps stiffness is an idealization that cannot exist in reality, in which case the cosmological constant is an impossibility. Quintessence with w near −1 may be the closest reasonable approximation.

Quintessence on the Brane

Saying that quintessence is a field is just the first step in explaining it. Where would such a strange field come from? Particle

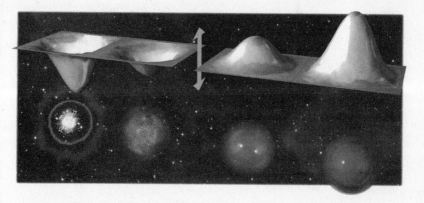

The missing matter that baffles the calculations of today's cosmologists may be hidden in a cosmic "blinking" action where subatomic particles pop in an out of existence millions of times a second.

physicists have explanations for phenomena from the structure of atoms to the origin of mass, but quintessence is something of an orphan. Modern theories of elementary particles include many kinds of fields that might have the requisite behavior, but not enough is known about their kinetic and potential energy to say which, if any, could produce negative pressure today.

An exotic possibility is that quintessence springs from the physics of extra dimensions. Over the past few decades, theorists have been exploring string theory, which may combine general relativity and quantum mechanics in a unified theory of fundamental forces. An important feature of string models is that they predict 10 dimensions. Four of these are our familiar three spatial dimensions, plus time. The remaining six must be hidden. In some formulations, they are curled up like a ball whose radius is too small to be detected (at least with present instruments). An alternative idea is found in a recent extension of string theory, known as M-theory, which adds an 11th dimension: ordinary matter is confined to two three-dimensional surfaces, known as branes (short for membranes), separated by a microscopic gap along the 11th dimension.

We are unable to see the extra dimensions, but if they exist, we should be able to perceive them indirectly. In fact, the presence of curled up dimensions or nearby branes would act just like a field. The numerical value that the field assigns to each point in space could correspond to the radius or gap distance. If the radius or gap changes slowly as the universe expands, it could exactly mimic the hypothetical quintessence field.

What a Coincidence

Whatever the origin of quintessence, its dynamism could solve the thorny problem of fine tuning. One way to look at this issue is to ask, Why has cosmic acceleration begun at this particular moment in cosmic history? Created when the universe was 10^{-35} seconds old, dark energy must have remained in the shadows for nearly 10 billion years—a factor of more than 10^{50} in age. Only then, the data suggest, did it overtake matter and cause the universe to begin accelerating. Is it not a coincidence that, just when thinking beings evolved, the universe suddenly shifted into overdrive? Somehow the fates of matter and of dark energy seem to be intertwined. But how?

If the dark energy is vacuum energy, the coincidence is almost impossible to account for. Some researchers, including Martin Rees of the University of Cambridge and Steven Weinberg of the University of Texas at Austin, have pursued an anthropic explanation. Perhaps our universe is just one among a multitude of universes, in each of which the vacuum energy takes on a different value. Universes with vacuum energy much greater than four electron volts per cubic millimeter might be more common, but they expand too rapidly to form stars, planets or life. Universes with much smaller values might be very rare. Our universe would have the optimal value. Only in this "best of all worlds" could there exist intelligent beings capable of contemplating the nature of the universe. But physicists disagree whether the anthropic argument constitutes an acceptable explanation.

A more satisfying answer, which could involve a form of quintessence known as a tracker field, was studied by Ratra and Peebles and by Steinhardt, Ivaylo Zlatev and Limin Wang of the University of Pennsylvania. The equations that describe tracker fields have classical attractor behavior like that found in some chaotic systems. In such systems, motion converges to the same result for a wide range of initial conditions. A marble put into an empty bathtub, for example, ultimately falls into the drain whatever its starting place.

Similarly, the initial energy density of the tracker field does not have to be tuned to a certain value, because the field rapidly adjusts itself to that value. It locks into a track on which its energy density remains a nearly constant fraction of the density of radiation and matter. In this sense, quintessence imitates matter and radiation, even though its composition is wholly different. The mimicking occurs because the radiation and matter density determine the cosmic expansion rate, which, in turn, controls the rate at which the quintessence density changes. On closer inspection, one finds that the fraction is slowly growing. Only after many millions or billions of years does quintessence catch up.

So why did quintessence catch up when it did? Cosmic acceleration could just as easily have commenced in the distant past or in the far future, depending on the choices of constants in the tracker field theory. This brings us back to the coincidence. But perhaps some event in the relatively recent past unleashed the acceleration. Steinhardt, along with Christian Armendáriz Picon and Viatcheslav Mukhanov of the Ludwig Maximilians University in Munich, has proposed one such recent event: the transition from radiation domination to matter domination.

According to the big bang theory, the energy of the universe used to reside mainly in radiation. As the universe cooled, however, the radiation lost energy faster than ordinary matter did. By the time the universe was a few tens of thousands of years old—a relatively short time ago in logarithmic terms—the energy balance had shifted in favor of matter. This change marked the beginning of the matter dominated epoch of which we are the beneficiaries. Only then could gravity begin to pull matter together to form galaxies and larger scale structures. At the same time, the expansion rate of the universe underwent a change.

In a variation on the tracker models, this transformation triggered a series of events that led to cosmic acceleration today. Throughout most of the history of the universe, quintessence tracked the radiation energy, remaining an insignificant component of the cosmos. But when the universe became matter dominated, the change in the expansion rate jolted quintessence out of its copycat behavior. Instead of tracking the radiation or even the matter, the pressure of quintessence switched to a negative value. Its density held nearly fixed and ultimately overtook the decreasing matter density. In this picture, the fact that thinking beings and cosmic acceleration came into existence at nearly the same time is not a coincidence. Both the formation of stars and planets necessary to support life and the transformation of quintessence into a negative pressure component were triggered by the onset of matter domination.

Looking to the Future

In the short term, the focus of cosmologists will be to detect the existence of quintessence. It has observable consequences. Because its value of w differs from that of vacuum energy, it produces a different rate of cosmic acceleration. More precise measurements of supernovae over a longer span of distances may separate the two cases. Astronomers have proposed two new observatories—the orbiting Supernova Acceleration Probe and the Earth-based Large Aperture Synoptic Survey Telescope—to resolve the issue. Differences in acceleration rate also produce small differences in the angular size of hot and cold spots in the cosmic microwave background radiation, as the Microwave Anisotropy Probe and Planck spacecraft should be able to detect.

Other tests measure how the number of galaxies varies with increasing redshift to infer how the expansion rate of the universe has changed with time. A ground based project known as the Deep Extragalactic Evolutionary Probe will look for this effect.

Over the longer term, all of us will be left to ponder the profound implications of these revolutionary discoveries. They lead to a sobering new interpretation of our place in cosmic history. In the beginning (or at least the earliest for which we have any clue), there was inflation, an extended period of accelerated expansion during the first instants after the big bang. Space back then was nearly devoid of matter, and a quintessence-like quantum field with negative pressure held sway. During that period, the universe expanded by a greater factor than it has during the 15 billion years since inflation ended. At the end of inflation, the field decayed to a hot gas of quarks, glutons, electrons, light and dark energy.

For thousands of years, space was so thick with radiation that atoms, let alone larger structures, could never form. Then matter took control. The next stage—our epoch—has been one of steady cooling, condensation and the evolution of intricate structure of ever increasing size. But this period is coming to an end. Cosmic acceleration is back. The universe as we know it, with shining stars, galaxies and clusters, appears to have been a brief interlude. As acceleration takes hold over the next tens of billions of years, the matter and energy in the universe will become more and more diluted and space will stretch too rapidly to enable new structures to form. Living things will find the cosmos increasingly hostile [see "The Fate of Life in the Universe," page 117]. If the acceleration is caused by vacuum energy, then the cosmic story is complete: the planets, stars and galaxies we see today are the pinnacle of cosmic evolution.

But if the acceleration is caused by quintessence, the ending has yet to be written. The universe might accelerate forever, or the quintessence could decay into new forms of matter and radiation, repopulating the universe. Because the dark energy density is so small, one might suppose that the material derived from its decay would have too little energy to do anything of interest. Under some circumstances, however, quintessence could decay through the nucleation of bubbles. The bubble interior would be a void, but the bubble wall would be the site of vigorous activity. As the wall moved outward, it would sweep up all the energy derived from the decay of quintessence. Occasionally, two bubbles would collide in a fantastic fireworks display. In the process, massive particles such as protons and neutrons might arise—perhaps stars and planets.

To future inhabitants, the universe would look highly inhomogeneous, with life confined to distant islands surrounded by vast voids. Would they ever figure out that their origin was the homogeneous and isotropic universe we see about us today? Would they ever know that the universe had once been alive and then died, only to be given a second chance?

Experiments may soon give us some idea which future is ours. Will it be the dead end of vacuum energy or the untapped potential of quintessence? Ultimately the answer depends on whether quintessence has a place in the basic workings of nature—the realm, perhaps, of string theory. Our place in cosmic history hinges on the interplay between the science of the very big and that of the very small.

"Where's the missing matter?" is one of the key questions that harries our current cosmological theorists. Our observations suggest a flat universe, where the outward push of cosmic expansion is balanced by the inward pull of gravity, producing a flat geometry. But even by the most generous estimates, the sum total of visible matter, radiation, and dark matter account for far less than one half of the matter predicted to exist in the universe.

Although no concrete solution yet exists, several intriguing theories have been suggested. One of the most curious candidates for the missing matter are structures called virtual particles. These invisible subatomic particles seem to appear from nothingness in a flash and then disappear again. It could be that the empty regions of space-time are more crowded than originally thought.

Cosmological Antigravity

Lawrence M. Krauss

N ovelist and social critic George Orwell wrote in 1946, "To see what is in front of one's nose requires a constant struggle." These words aptly describe the workings of modern cosmology. The universe is all around us—we are part of it—yet scientists must sometimes look halfway across it to understand the processes that led to our existence on the Earth. And although researchers believe that the underlying principles of nature are simple, unraveling them is another matter. The clues in the sky can be subtle. Orwell's adage is doubly true for cosmologists grappling with the recent observations of exploding stars hundreds of millions of light years away. Contrary to most expectations, they are finding that the expansion of the universe may not be slowing down but rather speeding up.

Astronomers have known that the visible universe is expanding since at least 1929, when Edwin P. Hubble demonstrated that distant galaxies are moving apart as they would if the entire cosmos were uniformly swelling in size. These outward motions are counteracted by the collective gravity of galaxy clusters and all the planets, stars, gas and dust they contain.

Even the minuscule gravitational pull of, say, a paper clip retards cosmic expansion by a slight amount. A decade ago a congruence of theory and observations suggested that there were enough paper clips and other matter in the universe to almost, but never quite, halt the expansion. In the geometric terms that Albert Einstein encouraged cosmologists to adopt, the universe seemed to be "flat."

That we live in a flat universe, the perfect balance of power, is one of the hallmark predictions of standard inflationary theory, which postulates a very early period of rapid expansion to reconcile several paradoxes in the conventional formulation of the big bang. Although the visible contents of the cosmos are clearly not enough to make the universe flat, celestial dynamics indicate that there is far more matter than meets the eye. Most of the material in galaxies and assemblages of galaxies must be invisible to telescopes. Over a decade ago I applied the term "quintessence" to this so-called dark matter, borrowing a term Aristotle used for the ether—the invisible material supposed to permeate all of space.

Yet an overwhelming body of evidence now implies that even the unseen matter is not enough to produce a flat universe. Perhaps the universe is not flat but rather open, in which case scientists must modify—or discard—inflationary theory. Or maybe the universe really is flat. If that is so, its main constituents cannot be visible matter, dark matter or radiation. Instead the universe must be composed largely of an even more ethereal form of energy that inhabits empty space, including that which is in front of our noses.

Fatal Attraction

The idea of such energy has a long and checkered history, which began when Einstein completed his general theory of relativity, more than a decade before Hubble's convincing demonstration that the universe is expanding. By tying together space, time and matter, relativity promised what had previously been impossible: a scientific understanding not merely of the dynamics of objects within the universe but of the universe itself. There was only one problem. Unlike other fundamental forces felt by matter, gravity is universally attractive—it only pulls; it cannot push. The unrelenting gravitational attraction of matter could cause the universe to collapse eventually. So Einstein, who presumed the universe to be static and stable, added an extra term to his equations, a "cosmological term," which could stabilize the universe by producing a new long range force throughout space. If its value were positive, the term would represent a repulsive force—a kind of antigravity that could hold the universe up under its own weight.

Physicists were happy to do without such an intrusion. In the general theory of relativity, the source of gravitational forces (whether attractive or repulsive) is energy. Matter is simply one form of energy. But Einstein's cosmological term is distinct. The energy associated with it does not depend on position or time—hence the name "cosmological constant." The force caused by the constant operates even in the complete absence of matter or radiation. Therefore, its source must be a curious energy that resides in empty space. The cosmological constant endows the void with an almost metaphysical aura. With its demise, nature was once again reasonable.

Or was it? In the 1930s glimmers of the cosmological constant arose in a completely independent context: the effort to combine the laws of quantum mechanics with Einstein's special theory of relativity. Physicists Paul A. M. Dirac and later Richard Feynman, Julian S. Schwinger and Shinichiro Tomonaga showed that empty space was more complicated than anyone had previously imagined. Elementary particles, it turned out, can spontaneously pop out of nothingness and disappear again, if they do so for a time so short that one cannot measure them directly. Such virtual particles, as they are called, may appear as far fetched as angels sitting on the head of a pin. But there is a difference. The unseen particles produce measurable effects, such as alterations to the energy levels of atoms as well as forces between nearby metal plates. The theory of virtual particles agrees with observations to nine decimal places. (Angels, in contrast, normally have no discernible effect on either atoms or plates.) Like it or not, empty space is not empty after all.

Virtual Reality

If virtual particles can change the properties of atoms, might they also affect the expansion of the universe? In 1967 Russian astrophysicist Yakov B. Zeldovich showed that the energy of virtual particles should act precisely as the energy associated with a cosmological constant. But there was a serious problem. Quantum theory predicts a whole spectrum of virtual particles, spanning every possible wavelength. When physicists add up all the effects, the total energy comes out infinite. Even if theorists ignore quantum effects smaller than a certain wavelength— for which poorly understood quantum gravitational effects presumably alter things—the calculated vacuum energy is roughly 120 orders of magnitude larger than the energy contained in all the matter in the universe.

What would be the effect of such a humongous cosmological constant? Taking a cue from Orwell's maxim, you can easily put an observational limit on its value. Hold out your hand and look at your fingers. If the constant were as large as quantum theory naively suggests, the space between your eyes and your hand would expand so rapidly that the light from your hand would never reach your eyes. To see what is in front of your face would be a constant struggle (so to speak), and you would always lose. The fact that you can see anything at all means that the energy of empty space cannot be large. And the fact that we can see not only to the ends of our arms but also to the far reaches of the universe puts an even more stringent limit on the cosmological constant: almost 120 orders of magnitude smaller than the estimate mentioned above. The discrepancy between theory and observation is the most perplexing quantitative puzzle in physics today. The simplest conclusion is that some as yet undiscovered physical law causes the cosmological constant to vanish. But as much as theorists might like the constant to go away, various astronomical observations—of the age of the universe, the density of matter and the nature of cosmic structures—all independently suggest that it may be here to stay.

Determining the age of the universe is one of the long-standing issues of modern cosmology. By measuring the velocities of galaxies, astronomers can calculate how long it took them to arrive at their present positions, assuming they all started out at the same place. For a first approximation, one can ignore the deceleration caused by gravity. Then the universe would expand at a constant speed and the time interval would just be the ratio of the distance between galaxies to their measured speed of separation—that is, the reciprocal of the famous Hubble constant. The higher the value of the Hubble constant, the faster the expansion rate and hence the younger the universe is.

Over the past seven decades, astronomers have improved their determination of the expansion rate, but the tension between the calculated age of the universe and the age of objects within it has persisted. In the past decade, with the launch of the Hubble Space Telescope and the development of new observational techniques, disparate measurements of the Hubble constant are finally beginning to converge. Wendy L. Freedman of the Carnegie Observatories and her colleagues have inferred a value of 73 kilometers per second per megaparsec (with a most likely range, depending on experimental error, of 65 to 81). These results put the upper limit on the age of a flat universe at about 10 billion years.

The Age Crisis

Is that value old enough? It depends on the age of the oldest objects that astronomers can date. Among the most ancient stars in our galaxy are those found in tight groups known as globular clusters, some of which are located in the outskirts of our galaxy and are thus thought to have formed before the rest of the Milky Way. Estimates of their age, based on calculations of how fast stars burn their nuclear fuel, traditionally ranged from 15 to 20 billion years. Such objects appeared to be older than the universe.

To determine whether this age conflict was the fault of cosmology or of stellar modeling, in 1995 my colleagues—Brian C. Chaboyer, then at the Canadian Institute of Theoretical Astrophysics, Pierre Demarque of Yale University and Peter J. Kernan of Case Western Reserve University—and I reassessed the globular cluster ages. We simulated the life cycles of three million different stars whose properties spanned the existing uncertainties, and then compared our model stars with those in globular clusters. The oldest, we concluded, could be as young as 12.5 billion years old, which was still at odds with the age of a flat, matter dominated universe.

But in 1997, the Hipparcos satellite, launched by the European Space Agency to measure the locations of over 100,000 nearby stars, revised the distances to these stars and, indirectly, to globular clusters. The new distances affected estimates of their brightness and forced us to redo our analysis, because brightness determines the rate at which stars consume fuel and hence their life spans. Now it seems that globulars could, at the limit of the observational error bars, be as young as 10 billion years old, which is just consistent with the cosmological ages.

But this marginal agreement is uncomfortable, because it requires that both sets of age estimates be near the edge of their allowed ranges. The only thing left that can give is the assumption that we live in a flat, matter-dominated universe. A lower density of matter, signifying an open universe with slower deceleration, would ease the tension somewhat. Even so, the only way to lift the age above 12.5 billion years would be to consider a universe dominated not by matter but by a cosmological constant. The resulting repulsive force would cause the Hubble expansion to accelerate over time. Galaxies would have been moving apart slower than they are today, taking longer to reach their present separation, so the universe would be older.

The current estimates of age are merely suggestive. Meanwhile other pillars of observational cosmology have recently been shaken, too. As astronomers have surveyed ever-larger regions of the cosmos, their ability to tally up its contents has improved. Now the case is compelling that the total amount of matter is insufficient to yield a flat universe.

This cosmic census first involves calculations of the synthesis of elements by the big bang. The light elements in the cosmos—hydrogen and helium and their rarer isotopes, such as deuterium—were created in the early universe in relative amounts that depended on the number of available protons and neutrons, the constituents of normal matter. Thus, by comparing the abundances of the various isotopes, astronomers can deduce the total amount of ordinary matter that was produced in the big bang. (There could, of course, also be other matter not composed of protons and neutrons.)

The relevant observations took a big step forward in 1996 when David R. Tytler and Scott Burles of the University of California at San Diego and their colleagues measured the primordial abundance of deuterium using absorption of quasar light by intergalactic hydrogen clouds. Because these clouds have never contained stars, their deuterium could only have been created by the big bang. Tytler and Burles's finding implies that the average density of ordinary matter is between four and seven percent of the amount needed for the universe to be flat.

Astronomers have also probed the density of matter by studying the largest gravitationally bound objects in the universe: clusters of galaxies. These groupings of hundreds of galaxies account for almost all visible matter. Most of their luminous content takes the form of hot intergalactic gas, which emits X rays. The temperature of this gas, inferred from the spectrum of the X rays, depends on the total mass of the cluster: in more massive clusters, the gravity is stronger and hence the pressure that supports the gas against gravity must be larger, which drives the temperature higher. In 1993 Simon D. M. White, now at the Max Planck Institute for Astrophysics in Garching, Germany, and his colleagues compiled information about several different clusters to argue that luminous matter accounted for between 10 and 20 percent of the total mass of the objects. When combined with the measurements of deuterium, these results imply that the total density of clustered matter—including protons and neutrons as well as more exotic particles such as certain dark-matter candidates—is at most 60 percent of that required to flatten the universe.

A third set of observations, ones that also bear on the distribution of matter at the largest scales, supports the view that the universe has too little mass to make it flat. Perhaps no other subfield of cosmology has advanced so much in the past 20 years as the understanding of the origin and nature of cosmic structures. Astronomers had long assumed that galaxies coalesced from slight concentrations of matter in the early universe, but no one knew what would have produced such undulations. The development of the inflationary theory in the 1980s provided the first plausible mechanism—namely, the enlargement of quantum fluctuations to macroscopic size.

Numerical simulations of the growth of structures following inflation have shown that if dark matter was not made from protons and neutrons but from some other type of particle, tiny ripples in the cosmic microwave background radiation could grow into the structures now seen. Moreover, concentrations of matter should still be evolving into clusters of galaxies if the overall density of matter is high. The relatively slow growth of the number of rich clusters over the recent history of the universe suggests that the density of matter is less than 50 percent of that required for a flat universe.

Nothing Matters

These many findings that the universe has too little matter to make it flat have become convincing enough to overcome the strong theoretical prejudice against this possibility. Two interpretations are viable: either the universe is open, or it is made flat by some additional form of energy that is not associated with ordinary matter. To distinguish between these alternatives, astronomers have been pushing to measure the microwave background at high resolution. Meanwhile researchers studying distant supernovae have provided the first direct, if tentative, evidence that the expansion of the universe is accelerating, a telltale sign of a cosmological constant with the same value implied by the other data. Observations of the microwave background and of supernovae illuminate two different aspects of cosmology. The microwave background reveals the geometry of the universe, which is sensitive to the total density of energy, in whatever form, whereas the supernovae directly probe the expansion rate of the universe, which depends on the difference between the density of matter (which slows the expansion) and the cosmological constant (which can speed it up).

Together all these results suggest that the constant contributes between 40 and 70 percent of the energy needed to make the universe flat. Despite the preponderance of evidence, it is worth remembering the old saw that an astronomical theory whose predictions agree with all observations is probably wrong, if only because some of the measurements or some of the predictions are likely to be erroneous. Nevertheless, theorists are already scrambling to understand what 20 years ago would have been unthinkable: a cosmological constant greater than zero yet much smaller than current quantum theories predict. Some feat of fine tuning must subtract virtual particle energies to 123 decimal places but leave the 124th untouched—a precision seen nowhere else in nature.

One direction, explored recently by Steven Weinberg of the University of Texas at Austin and his colleagues, invokes the last resort of cosmologists, the anthropic principle. If the observed universe is merely one of an infinity of disconnected universes—each of which might have slightly different constants of nature, as suggested by some incarnations of inflationary theory combined with emerging ideas of quantum gravity— then physicists can hope to estimate the magnitude of the cosmological constant by asking in which universes intelligent life is likely to evolve. Weinberg and others have arrived at a result that is compatible with the apparent magnitude of the cosmological constant today.

Most theorists, however, do not find these notions convincing, as they imply that there is no reason for the constant to take on a particular value; it just does. Although that argument may turn out to be true, physicists have not yet exhausted the other possibilities, which might allow the constant to be constrained by fundamental theory rather than by accidents of history.

Another direction of research follows in a tradition established by Dirac. He argued that there is one measured large number in the universe—its age (or, equivalently, its size). If certain physical quantities were changing over time, they might naturally be either very large or very small today. The cosmological constant could be one example. It might not, in fact, be constant. After all, if the cosmological constant is fixed and nonzero, we are living at the first and only time in the cosmic history when the density of matter, which decreases as the universe expands, is comparable to the energy stored in empty space. Why the coincidence? Several groups have instead imagined that some form of cosmic energy mimics a cosmological constant but varies with time.

This concept was explored by P. James E. Peebles and Bharat V. Ratra of Princeton University a decade ago. Motivated by the new supernova findings, other groups have resurrected the idea. Some have drawn on emerging concepts from string theory. Robert Caldwell and Paul J. Steinhardt of the University of Pennsylvania have reproposed the term "quintessence" to describe this variable energy. It is one measure of the theoretical conundrum that the dark matter that originally deserved this term now seems almost mundane by comparison. As much as I like the word, none of the theoretical ideas for this quintessence seems compelling. Each is ad hoc. The enormity of the cosmological constant problem remains.

How will cosmologists know for certain whether they have to reconcile themselves to this theoretically perplexing universe? New measurements of the microwave background, the continued analysis of distant supernovae and measurements of gravitational lensing of distant quasars should be able to pin down the cosmological constant over the next few years. One thing is already certain. The standard cosmology of the 1980s, postulating a flat universe dominated by matter, is dead. The universe is either open or filled with an energy of unknown origin. Although I believe the evidence points in favor of the latter, either scenario will require a dramatic new understanding of physics. Put another way, "nothing" could not possibly be more interesting.

Quantum mechanics describes the seemingly bizarre behavior of atomic and subatomic particles, which act in ways that defy our intuitive understanding of the world and the objects in it. For example, within the dictates of quantum mechanics, the particles that compose matter can be thought of as particles or as energy waves. Particles can pop in and out of physical existence. And they can encompass and materialize a literally infinite number of probabilities.

When quantum mechanics turns its gaze to the universe at large it becomes quantum cosmology. The nature of this line of thought becomes especially valuable when we approach the question of where the universe comes from. Big bang and inflation theories outline the "how" of the universe's birth, but quantum cosmology attempts to tackle the "why."

Quantum Cosmology and the Creation of the Universe

Jonathan J. Halliwell

Using Einstein's theory of general relativity to extrapolate back in time, investigators deduced that the universe emerged from a single, unbelievably small, dense, hot region. The events that have unfolded since that moment, including the formation of matter as well as its coalescence into galaxies, stars, planets and chemical systems, appear to be adequately described by conventional cosmology.

Yet the conventional ideas are incomplete. They fail to explain or even describe the ultimate origin of the universe. The most extreme extrapolation backward in time takes the universe down to a size at which it is necessary to incorporate that other great vision of modern physics: quantum theory. But the marriage of quantum theory and general relativity has been described as, at best, a shotgun wedding. Its consummation remains one of the outstanding problems of physics.

In recent decades, workers have begun to make some progress in applying quantum theory to the universe. These early steps have been promising enough to encourage those taking them to coin a name for their endeavor: quantum cosmology. Quantum cosmologists build on foundations laid in the 1960s by Bryce S. DeWitt of the University of Texas at Austin, Charles W. Misner of the University of Maryland and John A. Wheeler of Princeton University. Their studies mapped out how quantum mechanics might be applied to the entire universe. But the work was not taken very seriously until the 1980s, after classical theories of cosmology began to falter in their attempts to explain fully the beginning of the universe.

Most notable among the investigators who were drawn to this work are James B. Hartle of the University of California at Santa Barbara, Stephen W. Hawking of the University of Cambridge, Andrei D. Linde of the Lebedev Physical Institute in Moscow and Alexander Vilenkin of Tufts University. They put forward quite definite laws of initial conditions, that is, conditions that must have existed at the very moment of creation. When adjoined with suitable laws governing the evolution of the universe, such proposals could conceivably lead to a complete explanation of all cosmological observations and would therefore resolve important problems that plague the foundations of conventional cosmology.

The hot big bang model makes definite predictions about the universe as it exists now. It predicts the formation of nuclei, the relative abundances of certain elements and the existence and exact temperature of the microwave background—the glow of radiation left over from the initial explosion, which permeates the universe.

Despite its successes, the hot big bang model leaves many features of the universe unexplained. For example, the universe today includes a vast number of regions that in the hot big bang model could never have been in casual contact at any stage in their entire history. These regions are moving away from one another at such a rate that any information, even traveling at the speed of light, could not cover the distance between them. This "horizon problem" makes it difficult to account for the striking uniformity of the cosmic background radiation.

Then there is the "flatness problem." The hot big bang model indicates the universe to become more curved as time passes. But observations reveal that the spatial geometry of the part of the universe we can observe is extremely flat. The universe could exhibit such flatness only if it started out almost exactly flat—to within one part in 10^{60}. Many cosmologists consider such fine tuning deeply unnatural.

Perhaps most significant, the hot big bang model does not adequately explain the origin of large-scale structures, such as galaxies. Researchers, among them Edward R. Harrison of the University of Massachusetts at Amherst and Yakov B. Zeldovich of the Institute of Physical Problems in Moscow, offered partial explanations, showing how large-scale structures might appear from small fluctuations in the density of matter in an otherwise homogeneous early universe. But the fundamental origin of these fluctuations remained completely unknown. They had to be assumed as initial conditions.

In brief, therefore, the hot big bang model suffered from extreme dependence on initial conditions. Finding the present universe in this model would be as unlikely as finding a pencil balanced on its point after an earthquake.

Inflation improves dramatically on the hot big bang model in that it allows for the currently observed state of the universe to have arisen from a much broader, far more plausible set of initial conditions.[See "The Self-Reproducing Inflationary Universe," page 32]. Nevertheless, inflation does not relieve the observed state of the universe of all dependence on assumptions about initial conditions. In particular, inflation itself depends on a number of assumptions. For example, it would have occurred only if the scalar field began with a large, approximately constant energy density. This approximately constant energy density is equivalent, at least for a brief time, to Einstein's famous (or infamous) cosmological constant. Therefore, like it or not, the success of inflation rests on certain assumptions about initial conditions.

From where do these assumptions come? Obviously, one can go on asking an infinite sequence of such questions, like an overbearingly curious child in the "Why?" stage. But the cosmologist seeking a complete explanation is ultimately compelled to ask, "What happened before inflation? How did the universe actually begin?"

One can start answering these questions by following the expansion of the universe backward in time to the pre-inflation era. There the size of the universe tends to zero, and the strength of the gravitational field and the energy density of matter tend to infinity. That is, the universe appears to have emerged from a singularity, a region of infinite curvature and energy density at which the known laws of physics break down.

Singularities are not artifacts of the models. These conditions are a consequence of the famous "singularity theorems," proved in the 1960s by Hawking and Roger Penrose of the University of Oxford. These theorems showed that under reasonable assumptions any model of the expanding universe extrapolated backward in time will encounter an initial singularity.

The theorems do not imply, however, that a singularity will physically occur. Rather the theory predicting them—classical general relativity—breaks down at very high curvatures and must be superseded by some bigger, better, more powerful theory. What is this theory? A consideration of scale yields a clue. Near a singularity, space-time becomes highly curved; its volume shrinks to very small dimensions. Under such circumstances, one must appeal to the theory of the very small—that is, to quantum theory.

Quantum theory arose from an attempt to explain phenomena that lay beyond the scope of conventional classical physics. A central failure of classical mechanics was its inability to account for the structure of the atom. Experiments suggested that the atom consisted of electrons orbiting a nucleus, much as planets orbit the sun. Efforts to describe this model using classical physics, however, predicted that the electrons should plunge into the nucleus. There was nothing to hold them in orbit.

To overcome the discrepancy between observation and theory, Niels Bohr, Erwin Schrödinger, Werner K. Heisenberg and Paul A. M. Dirac, among others, in the early 20th century developed quantum mechanics. In this formulation, motion is not deterministic (as in classical mechanics) but probabilistic. The dynamic variables of classical mechanics, such as position and momentum, do not in general have definite values in quantum mechanics, which regards a system as fundamentally wavelike in nature. A quantity called the wave function encodes the probabilistic information about such variables as position, momentum and energy. One finds the wave function for a system by solving an equation called the Schrödinger equation.

For a single-point particle, one can regard the wave function as an oscillating field spread throughout physical space. At each point in space, the function has an amplitude and a wavelength. The square of the amplitude is proportional to the probability of finding the particle at that position. For wave functions that have constant amplitudes, the wavelength is related to the momentum of the particle. But because the wave functions for position and momentum are mutually exclusive, an indefiniteness, or uncertainty, in both quantities will always exist. As the measurement of one property, say, position, becomes more precise, the value for the other grows correspondingly indefinite. This state of affairs, called Heisenberg's uncertainty principle, is an elementary consequence of the wavelike nature of particles.

The uncertainty principle leads to phenomena qualitatively different from those exhibited in classical mechanics. In quantum mechanics, a system can never have an energy of exactly zero. The total energy is generally the sum of the kinetic and potential energies. The kinetic energy depends on momentum;

potential energy depends on position (a ball on top of a hill has more gravitational potential energy than one that sits in a well). Because the uncertainty principle forbids any simultaneous, definite values of momentum and position, the kinetic and potential energy cannot both be exactly zero.

Instead the system has a ground state in which the energy is as low as it can be. (Recall that in the inflationary universe scenario, galaxies form from "ground-state fluctuations.") Such fluctuations also prevent the orbiting electron from crashing into the nucleus. The electrons have an orbit of minimum energy from which they cannot fall into the nucleus without violating the uncertainty principle.

Uncertainty also leads to the phenomenon of tunneling. In classical mechanics, a particle traveling with fixed energy cannot penetrate an energy barrier. A ball at rest in a bowl will never be able to get out. In quantum mechanics, position is not sharply defined but is spread over a (typically infinite) range. As a result, there is a definite probability that the particle will be found on the other side of the barrier. One says that the particle can "tunnel" through the barrier.

The tunneling process should not be thought of as occurring in real time. In a certain well-defined mathematical sense, the particle is conveniently thought of as penetrating the barrier in "imaginary" time, that is, time multiplied by the square root of minus one. (Time here loses its meaning in the usual sense of the word; it actually resembles a spatial dimension more than it does real time.)

These distinctly quantum effects do not contradict classical mechanics. Rather quantum mechanics is a broader theory and supersedes classical mechanics as the correct description of nature. On macroscopic scales, the wavelike nature of particles is highly suppressed, so that quantum mechanics reproduces the effects of classical mechanics to a high degree of precision (although how this "quantum to classical" transition comes about is still a matter of current research).

*   *   *

How can these insights be employed to illuminate questions of cosmology? Like quantum mechanics, quantum cosmology attempts to describe a system fundamentally in terms of its wave function. One can find the wave function of the universe by solving an equation called the Wheeler-DeWitt equation, which is the cosmological analogue of the Schrödinger equation. In the simplest cases, the spatial size of the universe is the analogue of position, and the rate of the universe's expansion represents the momentum.

Yet many conceptual and technical difficulties arise in quantum cosmology above and beyond those in quantum mechanics. The most serious is the lack of a complete, manageable quantum theory of gravity. Three of the four fundamental forces of nature—electromagnetism, the strong nuclear force and the weak nuclear force—have been made consistent with quantum theory. But all attempts to quantize Einstein's general relativity have met with failure. The failure looms large: recall that general relativity, the best theory of gravity that we have, says that at the singularity, space becomes infinitely small and the energy density infinitely great. To look beyond such a moment requires a quantum theory of gravity.

Another question that workers confront is the applicability of quantum mechanics to the entire universe. Quantum mechanics was developed to describe atomic-scale phenomena. The beautiful agreement between quantum mechanics and experiment is one of the great triumphs of modern physics; no physicist in his or her right mind harbors any doubts as to its correctness on the atomic scale. But a few may raise dissenting voices if one suggests that quantum mechanics is equally applicable to, say, tables and chairs.

The challenge is not easy to dismiss, because on the macroscopic scale the predictions of quantum mechanics coincide closely with those of classical mechanics. Genuine macroscopic quantum effects are extremely difficult to detect experimentally. Even more contentious is the most extravagant extrapolation possible: that quantum mechanics applies to the entire universe at all times and to everything in it. Acceptable or not, this is the fundamental assertion of quantum cosmology.

Another, perhaps more difficult issue concerns the interpretation of quantum mechanics applied to cosmology. In the development of quantum mechanics (as applied to atoms), it proved necessary to understand how the mathematics of the theory translates into what one would actually observe during a measurement. Bohr laid the foundations of this translation, known as quantum measurement theory, in the 1920s and 1930s. He assumed that the world may be divided into two parts: microscopic systems (such as atoms), governed purely by quantum mechanics, and external macroscopic systems (such as observers and their measuring apparata), governed by classical mechanics. A measurement is an interaction between the observer and the microscopic system that leads to a permanent recording of the event.

During this interaction, the wave function describing the microscopic system undergoes a discontinuous change from its initial state to some final one. The quantity being measured takes on a definite value in the final state. The discontinuous change is referred to, rather dramatically, as the collapse of the wave function. For instance, the wave function could start out in a state of definite momentum, but if position is being measured, it "collapses" into a state of definite position.

Although many theorists feel this scheme, known as the Copenhagen interpretation of quantum mechanics, is philosophically unsatisfactory, it nonetheless enables predictions to be extracted from theory—predictions that agree with observation. Perhaps for this reason, the Copenhagen interpretation has stood largely unchallenged for almost half a century.

In attempting to apply quantum mechanics to the entire universe, however, one meets with acute difficulties that cannot be brushed off as philosophical niceties. In a theory of the universe, of which the observer is a part, there should be no fundamental division between observer and observed. Moreover, most researchers feel uncomfortable at the thought of the wave function of the entire universe collapsing when an observation is made. Questions concerning probability predictions also come up. Ordinarily, one tests such predictions by making a large number of measurements. For example, flipping a coin many times will verify that the probability of heads is one half. In cosmology, there is only one system, which is measured only once.

Keeping such difficulties in mind, Hugh Everett III of Princeton, one of the first physicists to take seriously the notion of applying quantum mechanics to the universe, presented a framework for the interpretation of quantum mechanics particularly suited to the special needs of cosmology. Unlike Bohr, Everett asserted that there exists a universal wave function describing both macroscopic observers and microscopic systems, with no fundamental division between them. A measurement is just an interaction between different parts of the entire universe, and the wave function should predict what one part of the system "sees" when it observes another.

Hence, there is no collapse of the wave function in Everett's picture, only a smooth evolution described by the Schrödinger equation for the entire system. But as he modeled the measurement process, Everett made a truly remarkable discovery: the measurement appears to cause the universe to "split" into sufficiently many copies of itself to take into account all possible outcomes of the measurement.

Theorists have hotly debated the reality of the multiple copies in Everett's uneconomical "many worlds" interpretation, indeed, modern versions of Everett's idea, generated most notably by Murray Gell-Mann of the California Institute of Technology and Hartle, play down the many-worlds aspect of the theory. Instead their versions talk about "decoherent histories," which are possible histories for the universe to which probabilities may be assigned. For practical purposes, it does not matter whether one thinks of all or just one of them as actually happening. These ideas also have the great merit of eliminating the role of the observer and the need to collapse the wave function. And despite the controversy, such approaches give theorists some kind of framework within which to work.

Gell-Mann and Hartle also address the issue of probabilities for the universe. They insist that the only probabilities that have any meaning in quantum cosmology are a priori ones. These probabilities are close to one or zero, that is, definite yes-no predictions. Although most probabilistic predictions are not of this type, they can often be made so by suitably modifying the questions one asks. Unlike quantum mechanics, in which the goal is to determine probabilities for the possible outcomes of given observations, quantum cosmology seeks to determine those observations for which the theory gives probabilities close to zero or one.

This kind of approach has led to the following understanding: at certain points in space and time, typically (but not always) when the universe is large, the wave function for the universe indicates that the universe behaves classically to a high degree of precision. Classical space-time is then a prediction of the theory. Under these circumstances, moreover, the wave function provides probabilities for the set of possible classical behaviors of the universe.

On the other hand, certain regions, such as those close to classical singularities, exist in which no such prediction is possible. There the notions of space and time quite simply do not exist. There is just a "quantum fuzz," still describable by known laws of quantum physics but not by classical laws. Hence, in quantum cosmology, one no longer worries about trying to impose classical initial conditions on a region in which classical physics is not valid, such as near the initial singularity.

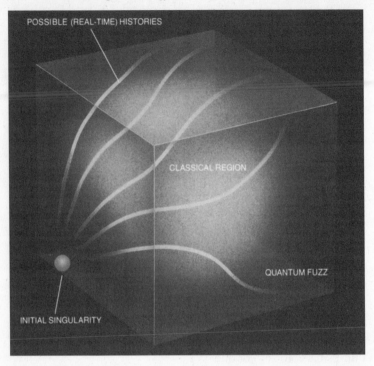

POSSIBLE (REAL-TIME) HISTORIES

CLASSICAL REGION

QUANTUM FUZZ

INITIAL SINGULARITY

From the point of initial singularity, the moment of the universe's birth, quantum theory holds that all possible universes are not only possible, but they exist and are as valid as our own universe. The lines running through this box representing "quantum fuzz" show some of the probable paths our universe might have taken.

Still, the wave function of the universe described by the quantum theory of cosmology does not eliminate the need for assumed initial conditions. Instead the question of classical initial conditions—the assumptions of the inflation and big bang models—becomes one of quantum initial conditions: Of the many wave functions possible (the many solutions to the Wheeler-DeWitt equation), how is just one singled out?

The problem is best understood by contrasting the cosmological situation with that of the laboratory, to which most of science is directed. There a system has clearly defined temporal and spatial boundaries—the duration of the reaction, for instance, or the size of the beaker. At those boundaries, experimenters may control, or at least observe, the physical states. Using suitable laws of physics, they may be able to determine how the initial or boundary conditions evolve in space and time.

In cosmology the system under scrutiny is the entire universe. By definition, it has no exterior, no outside world, no "rest of the universe" to which one could appeal for boundary or initial conditions. Furthermore, it seems most unlikely that mathematical consistency alone will lead to a unique solution to the Wheeler-DeWitt equation, as DeWitt once suggested. Therefore, in much the same way that the theoretical physicist proposes laws to govern the evolution of physical systems, the inescapable task of the quantum cosmologist is to propose laws of initial or boundary conditions for the universe. In particular, Hartle and Hawking, Linde, and Vilenkin have made quite definite proposals that were intended to pick out a particular solution to the Wheeler-DeWitt equation, that is, to single out a unique wave function for the universe.

Hartle and Hawking's proposal defines a particular wave function of the universe using a rather elegant formulation of quantum mechanics originally developed in the 1940s by the late Richard P. Feynman of Caltech. The formulation is called the path integral or sum-over-histories method. In ordinary quantum mechanics, calculation of the wave function involves performing a certain sum over a class of histories of the system. The histories end at the point in space and time at which one wishes to know the value of the wave function. To render the wave function unique, one specifies precisely the class of histories to be summed over. The specified class includes not only classical histories but all possible histories for the system.

Summing over histories is mathematically equivalent to solving the Schrödinger equation. But it provides a very different view of quantum mechanics that has proved extremely useful, both technically and conceptually. In particular, the sum-over-histories method readily generalizes to quantum cosmology. The wave function of the universe may be calculated by summing over some class of histories for the universe. The technique is equivalent to solving the Wheeler-DeWitt equation, as was demonstrated most generally in a recent paper by Hartle and me. The precise solution obtained depends on how the class of histories summed over is chosen.

One way to understand the choice made by Hartle and Hawking is to translate their mathematics into geometry. Imagine the spatial extent of the universe at a particular time as a closed loop of string lying in the horizontal plane. If the vertical axis represents time, then the loop changes in size as time passes (representing the expansion and contraction of the universe). Various possible histories of the universe therefore appear as tubes swept out by the loop as it evolves in time. The final edge represents the universe today; the opposite end is the initial state (that is the creation of the universe), to be specified by proposed boundary conditions. Some tubes might close off in a sharp way, like the point of a cone; others might simply end abruptly.

Hartle and Hawking proposed that one should consider only tubes whose initial end shrinks to zero in a smooth, regular way, forming a kind of a hemispherical cap. One therefore sums over geometries that have no boundary (except for the final end, which is open and corresponds to the present universe). Hence, Hartle and Hawking's idea is called the no-boundary proposal.

Closing off the geometry in such a smooth way is impossible in classical theory. The singularity theorems imply that the classical histories of the universe must shrink to zero in a singular way, much as the end of a cone shrinks to a point. But in quantum theory the sum-over-histories approach admits many possible histories, not just classical ones.

This reasoning has given rise to another proposal, or solution, to the Wheeler-DeWitt equation. Recall that the appearance of imaginary time is characteristic of tunneling processes in quantum theory. Perhaps, then, the universe has tunneled from "nothing." The evolution described by inflation and the big bang would have subsequently occurred after the tunneling. The no-boundary wave function, however, does not have the general features normally associated with tunneling. It gives high probability of a classical universe appearing with a large size and a low energy density. An ordinary tunneling process would suppress a transition from zero to large size and give highest probability for tunneling to small size that has a high energy density.

Partly for this reason, Linde and Vilenkin independently put forward a "tunneling" proposal. The precise statement of this idea is mathematical, but it suffices to say that the scheme is designed to pick out a solution to the Wheeler-DeWitt equation possessing the properties expected of a tunneling process. Their solution enables one to think more appropriately of the universe as *tunneling from nothing*.

The no-boundary and tunneling proposals select a unique wave function for the universe (contingent, however, on the resolution of a number of technical difficulties recently exposed by Hartle, Jorma Louko of the University of Alberta and me). In both, the wave function indicates that space-time behaves according to classical cosmology when the universe is a few thousand times larger than the size at which the four forces of nature would be unified (about 10^{-33} centimeter), in agreement with observation. When the universe is smaller, however, the wave function indicates that classical space-time does not exist.

Given a unique wave function of the universe, one may finally ask, "How did the universe actually begin?" Rather than answering, a quantum cosmologist would reframe the question. In the neighborhood of singularities, the wave functions given by the tunneling and no-boundary proposals state that classical general relativity is not valid. Furthermore, the notions of space and time implicit in the question become inappropriate. The picture that emerges is of a universe with nonzero size and finite (rather than infinite) energy density appearing from a quantum fuzz.

After quantum creation, the wave function assigns probabilities to different evolutionary paths, one of which includes the inflation postulated by Guth. Although some theorists disagree, both the no-boundary and tunneling proposals seem to predict the conditions necessary for inflation, thereby eliminating the need for assumptions about the scalar-field matter that drove the rapid expansion.

The no-boundary and tunneling proposals also eliminate assumptions about the density perturbations. Although inflation explains their origin, the exact form and magnitude depend on certain assumptions about the initial state of the scalar-field matter. The inflation model assumes the inhomogeneous parts started out in their quantum mechanical ground state—the lowest possible energy state consistent with the uncertainty principle.

But in 1985 Hawking and I demonstrated this assumption must be a consequence of the no-boundary proposal: the correct kinds of inhomogeneities emerge naturally from the theory. The no-boundary proposal states that everything must be smooth and regular on the bottom cap of the space-time tube. This condition implies that inhomogeneous fluctuations must be zero there. Evolving up the tube in imaginary time, the fluctuations grow and enter the real-time region as small as they could possibly be—as the quantum mechanical ground-state fluctuations demanded by the inflation model. The tunneling proposal makes the same prediction, for similar reasons.

So we arrive at a possible answer. According to the picture afforded by quantum cosmology, the universe appeared from a quantum fuzz, tunneling into existence and thereafter evolving classically. The most compelling aspect of this picture is that the assumptions necessary for the inflationary universe scenario may be compressed into a single, simple boundary condition for the wave function of the universe.

How can one verify a law of initial conditions? An indirect test is to compare the predictions of the quantum models with the initial conditions needed for standard classical cosmological models. In this endeavor, as we have seen, quantum cosmologists can claim a reasonable degree of success.

More direct, observational tests are difficult. Much has happened in the universe since its birth, and each stage of evolution has to be modeled separately. It is difficult to distinguish between effects that result from a particular set of initial conditions and those that derive from the evolution of the universe or from the modeling of a particular stage.

What is needed is an observation of some effect that was produced at the beginning of the universe but was insensitive to the subsequent evolution. In 1987 Leonid Grishchuk of the Sternberg Astronomy Institute in Moscow argued that gravitational waves may be the sought-after effect. Quantum creation scenarios produce gravitational waves of a calculable form and magnitude. Gravitational waves interact very weakly with matter as they propagate through space-time. Therefore, when we observe them in the present universe, their spectrum may still contain the signature of quantum creation. Detecting gravitational waves is, unfortunately, extremely difficult, and current attempts have failed. New detectors to be built later this decade may prove sensitive enough to find the waves.

Because it is so hard to verify quantum cosmology, we cannot conclusively determine whether the no-boundary or the tunneling proposals are the correct ones for the wave function of the universe. It could be a very long time before we can tell if either is an answer to the question, "Where did all this come from?" Nevertheless, through quantum cosmology, we have at least been able to formulate and address the question in a meaningful—and most interesting—way.

Sometime in the future, our sun, Sol, will perish, and with it, life on Earth will perish too. Yet the boundless imagination and ingenuity of the human family refuses to accept such a fate as the end of our development. In a brilliant series of musings based on the ideas of Freeman Dyson, Lawrence Krauss and Glenn Starkman explore the possibilities and the probabilities of our species living forever.

The Fate of Life in the Universe

Lawrence M. Krauss and Glenn D. Starkman

Eternal life is a core belief of many of the world's religions. Usually it is extolled as a spiritual Valhalla, an existence without pain, death, worry or evil, a world removed from our physical reality. But there is another sort of eternal life that we hope for, one in the temporal realm. In the conclusion to *On The Origin of Species*, Charles Darwin wrote: "As all the living forms of life are the lineal descendants of those which lived before the Cambrian epoch, we may feel certain that the ordinary succession by generation has never once been broken. . . . Hence we may look with some confidence to a secure future of great length." The sun will eventually exhaust its hydrogen fuel, and life as we know it on our home planet will eventually end, but the human race is resilient. Our progeny will seek new homes, spreading into every corner of the universe just as organisms have colonized every possible niche of the Earth. Death and evil will take their toll, pain and worry may never go away, but somewhere we expect that some of our children will carry on.

Or maybe not. Remarkably, even though scientists fully

understand neither the physical basis of life nor the unfolding of the universe, they can make educated guesses about the destiny of living things. Cosmological observations now suggest the universe will continue to expand forever—rather than, as scientists once thought, expand to a maximum size and then shrink. Therefore, we are not doomed to perish in a fiery "big crunch" in which any vestige of our current or future civilization would be erased. At first glance, eternal expansion is cause for optimism. What could stop a sufficiently intelligent civilization from exploiting the endless resources to survive indefinitely?

Yet life thrives on energy and information, and very general scientific arguments hint that only a finite amount of energy and a finite amount of information can be amassed in even an infinite period. For life to persist, it would have to make do with dwindling resources and limited knowledge. We have concluded that no meaningful form of consciousness could exist forever under these conditions.

The Deserts of Vast Eternity

Over the past century, scientific eschatology has swung between optimism and pessimism. Not long after Darwin's confident prediction, Victorian-era scientists began to fret about the "heat death," in which the whole cosmos would come to a common temperature and thereafter be incapable of change. The discovery of the expansion of the universe in the 1920s allayed this concern, because expansion prevents the universe from reaching such an equilibrium. But few cosmologists thought through the other implications for life in an ever expanding universe, until a classic paper in 1979 by physicist Freeman Dyson of the Institute for Advanced Study in Princeton, N.J., itself motivated by earlier work by Jamal Islam, now at the University of Chittagong in Bangladesh. Since Dyson's paper, physicists and astronomers have periodically reexamined the topic. A year ago, spurred on by new observations that suggest a drastically different long-term future for the universe than that previously envisaged, we decided to take another look.

Over the past 12 billion years or so, the universe has passed through many stages. At the earliest times for which scientists now have empirical information, it was incredibly hot and dense. Gradually, it expanded and cooled. For hundreds of thousands of years, radiation ruled; the famous cosmic microwave background radiation is thought to be a vestige of this era. Then matter started to dominate, and progressively larger astronomical structures condensed out. Now, if recent cosmological observations are correct, the expansion of the

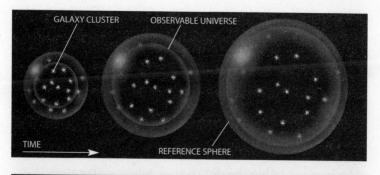

The same start, two different outcomes. Intelligent beings on a galaxy cluster inside a universe that is decelerating (top) would see more stars as their reference sphere grows; the opposite holds true if the universe is expanding (bottom).

universe is beginning to accelerate—a sign that a strange new type of energy, perhaps springing from space itself, may be taking over.

Life as we know it depends on stars. But stars inevitably die, and their birth rate has declined dramatically since an initial burst about 10 billion years ago. About 100 trillion years from now, the last conventionally formed star will wink out, and a new era will commence. Processes currently too slow to be noticed will become important: the dispersal of planetary systems by stellar close encounters, the possible decay of ordinary and exotic matter, the slow evaporation of black holes.

Assuming that intelligent life can adapt to the changing circumstances, what fundamental limits does it face? In an eternal universe, potentially of infinite volume, one might hope that a sufficiently advanced civilization could collect an infinite amount of matter, energy and information. Surprisingly, this is not true. Even after an eternity of hard and well-planned labor, living beings could accumulate only a finite number of particles, a finite quantity of energy and a finite number of bits of information. What makes this failure all the more frustrating is that the number of available particles, ergs and bits may grow without bound. The problem is not necessarily the lack of resources but rather the difficulty in collecting them.

The culprit is the very thing that allows us to contemplate an eternal tenure: the expansion of the universe. As the cosmos grows in size, the average density of ordinary sources of energy declines. Doubling the radius of the universe decreases the density of atoms eightfold. For light waves, the decline is even more precipitous. Their energy density drops by a factor of 16 because the expansion stretches them and thereby saps their energy.

As a result of this dilution, resources become ever more time consuming to collect. Intelligent beings have two distinct strategies: let the material come to them or try to chase it down. For the former, the best approach in the long run is to let gravity do the work. Of all the forces of nature, only gravity and electromagnetism can draw things in from arbitrarily far away. But the latter gets screened out: oppositely charged particles balance one another, so that the typical object is neutral and hence immune to long-range electrical and magnetic forces. Gravity, on the other hand, cannot be screened out, because particles of matter and radiation only attract gravitationally; they do not repel.

Surrender to the Void

Even gravity, however, must contend with the expansion of the universe, which pulls objects apart and thereby weakens their mutual attraction. In all but one scenario, gravity eventually becomes unable to pull together larger quantities of material. Indeed, our universe may have already reached this point; clusters of galaxies may be the largest bodies that gravity will ever be able to bind together. The lone exception occurs if the universe is poised between expansion and contraction, in which case gravity continues indefinitely to assemble ever greater amounts of matter. But that scenario is now thought to contradict observations, and in any event it poses its own difficulty: after 10^{33} years or so, the accessible matter will become so concentrated that most of it will collapse into black holes, sweeping up any lifeforms. Being inside a black hole is not a happy condition. On the earth, all roads may lead to Rome, but inside a black hole, all roads lead in a finite amount of time to the center of the hole, where death and dismemberment are certain.

Sadly, the strategy of actively seeking resources fares no better than the passive approach does. The expansion of the universe drains away kinetic energy, so prospectors would have to squander their booty to maintain their speed. Even in the most optimistic scenario—in which the energy is traveling toward the scavenger at the speed of light and is collected without

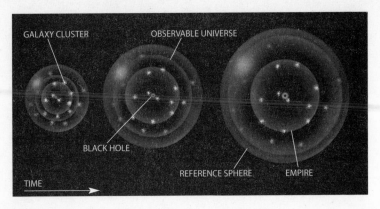

As a civilization expands to harvest energy and resources from other galaxy clusters, it grows and ultimately brings about its own downfall. Even tricks like squeezing energy out of black holes only go so far, as the "empire" finally reaches a size where further growth costs too much in terms of energy.

loss—a civilization could garner limitless energy only in or near a black hole. The latter possibility was explored by Steven Frautschi of the California Institute of Technology in 1982. He concluded that the energy available from the holes would dwindle more quickly than the costs of scavenging. We recently reexamined this possibility and found that the predicament is even worse than Frautschi thought. The size of a black hole required to sweep up energy forever exceeds the extent of the visible universe.

The cosmic dilution of energy is truly dire if the universe is expanding at an accelerating rate. All distant objects that are currently in view will eventually move away from us faster than the speed of light and, in doing so, disappear from view. The total resources at our disposal are therefore limited by what we can see today, at most.

Not all forms of energy are equally subject to the dilution. The universe might, for example, be filled with a network of cosmic strings—infinitely long, thin concentrations of energy that could have developed as the early universe cooled unevenly. The energy per unit length of a cosmic string remains unchanged despite cosmic expansion. Intelligent beings might try to cut one, congregate around the loose ends and begin consuming it. If the string network is infinite, they might hope to satisfy their appetite forever. The problem with this strategy is that whatever lifeforms can do, natural processes can also do. If a civilization can figure out a way to cut cosmic strings, then the string network will fall apart of its own accord. For example, black holes may spontaneously appear on the strings and devour them. Therefore, the beings could swallow only a finite amount of string before running into another loose end. The entire string network would eventually disappear, leaving the civilization destitute.

What about mining the quantum vacuum? After all, the cosmic acceleration may be driven by the so-called cosmological constant, a form of energy that does not dilute as the universe expands. If so, empty space is filled with a bizarre type of radiation, called Gibbons-Hawking or de Sitter radiation. Alas, it is impossible to extract energy from this radiation for useful work. If the vacuum yielded up energy, it would drop into a lower energy state, yet the vacuum is already the lowest energy state there is.

No matter how clever we try to be and how cooperative the universe is, we will someday have to confront the finiteness of the resources at our disposal. Even so, are there ways to cope forever?

The obvious strategy is to learn to make do with less, a scheme first discussed quantitatively by Dyson. In order to reduce energy consumption and keep it low despite exertion, we would eventually have to reduce our body temperature. One might speculate about genetically engineered humans who function at somewhat lower temperatures than 310 kelvins (98.6 degrees Fahrenheit). Yet the human body temperature cannot be reduced arbitrarily; the freezing point of blood is a firm lower limit. Ultimately, we will need to abandon our bodies entirely.

While futuristic, the idea of shedding our bodies presents no fundamental difficulties. It presumes only that consciousness is not tied to a particular set of organic molecules but rather can be embodied in a multitude of different forms, from cyborgs to sentient interstellar clouds. Most modern philosophers and cognitive scientists regard conscious thought as a process that a computer could perform. The details need not concern us here (which is convenient, as we are not competent to discuss them). We still have many billions of years to design new physical incarnations to which we will someday transfer our conscious selves. These new "bodies" will need to operate at cooler temperatures and at lower metabolic rates—that is, lower rates of energy consumption.

Dyson showed that if organisms could slow their metabo-

The Fate of the Universe

The cosmological constant changes the usual simple picture of the future of the universe. Traditionally, cosmology has predicted two possible outcomes that depend on the geometry of the universe or, equivalently, on the average density of matter. If the density of a matter filled universe exceeds a certain critical value, it is "closed," in which case it will eventually stop expanding, start contracting and ultimately vanish in a fiery apocalypse. If the density is less than the critical value, the universe is "open" and will expand forever. A "flat" universe, for which the density equals the critical value, also will expand forever but at an ever-lower rate. Yet these scenarios assume that the cosmological constant equals zero. If not, it—rather than matter—may control the ultimate fate of the universe. The reason is that the constant, by definition, represents a fixed density of energy in space. Matter cannot compete: a doubling in radius dilutes its density eightfold. In an expanding universe the energy density associated with a cosmological constant must win out. If the constant has a positive value, it generates a long range repulsive force in space, and the universe will continue to expand even if the total energy density in matter and in space exceeds the critical value. (Large negative values of the constant are ruled out because the resulting attractive force would already have brought the universe to an end.) Even this new prediction for eternal expansion assumes that the constant is indeed constant, as general relativity suggests that it should be. If in fact the energy density of empty space does vary with time, the fate of the universe will depend on how it does so. And there may be a precedent for such changes—namely, the inflationary expansion in the primordial universe. Perhaps the universe is just now entering a new era of inflation, one that may eventually come to an end.

—L.M.K.

lisms as the universe cooled, they could arrange to consume a finite total amount of energy over all of eternity. Although the lower temperatures would also slow consciousness—the number of thoughts per second—the rate would remain large enough for the total number of thoughts, in principle, to be unlimited. In short, intelligent beings could survive forever, not just in absolute time but also in subjective time. As long as organisms were guaranteed to have an infinite number of thoughts, they would not mind a languid pace of life. When billions of years stretch out before you, what's the rush?

At first glance, this might look like a case of something for nothing. But the mathematics of infinity can defy intuition. For an organism to maintain the same degree of complexity, Dyson argued, its rate of information processing must be directly proportional to body temperature, whereas the rate of energy consumption is proportional to the square of the temperature (the additional factor of temperature comes from basic thermodynamics). Therefore, the power requirements slacken faster than cognitive alacrity does. At 310 kelvins, the human body expends approximately 100 watts. At 155 kelvins, an equivalently complex organism could think at half the speed but consume a quarter of the power. The trade-off is acceptable because physical processes in the environment slow down at a similar rate.

To Sleep, to Die

Unfortunately, there is a catch. Most of the power is dissipated as heat, which must escape—usually by radiating away—if the object is not to heat up. Human skin, for example, glows in infrared light. At very low temperatures, the most efficient radiator would be a dilute gas of electrons. Yet the efficiency even of this optimal radiator declines as the cube of the temperature, faster than the decrease in the metabolic rate. A point would come when organisms could not lower their temperature further. They would be forced instead to reduce their complexity—to dumb down. Before long, they could no longer be regarded as intelligent.

To the timid, this might seem like the end. But to compensate for the inefficiency of radiators, Dyson boldly devised a strategy of hibernation. Organisms would spend only a small fraction of their time awake. While sleeping, their metabolic rates would drop, but—crucially—they would continue to dissipate heat. In this way, they could achieve an ever lower average body temperature. In fact, by spending an increasing fraction of their time asleep, they could consume a finite amount of energy yet exist forever and have an infinite number of thoughts. Dyson concluded that eternal life is indeed possible.

Since his original paper, several difficulties with his plan have emerged. For one, Dyson assumed that the average temperature of deep space—currently 2.7 kelvins, as set by the cosmic microwave background radiation—would always decrease as the cosmos expands, so that organisms could continue to decrease their temperature forever. But if the universe has a cosmological constant, the temperature has an absolute floor fixed by the Gibbons-Hawking radiation. For current estimates of the value of the cosmological constant, this radiation has an effective temperature of about 10^{-29} kelvin. As was pointed out independently by cosmologists J. Richard Gott II, John Barrow, Frank Tipler and us, once organisms had cooled to this level, they could not continue to lower their temperature in order to conserve energy.

The second difficulty is the need for alarm clocks to wake the organisms periodically. These clocks would have to operate reliably for longer and longer times on less and less energy. Quantum mechanics suggests that this is impossible. Consider, for example, an alarm clock that consists of two small balls that are taken far apart and then aimed at each other and released. When they collide, they ring a bell. To lengthen the time between alarms, organisms would release the balls at a slower speed. But eventually the clock will run up against constraints from Heisenberg's uncertainty principle, which prevents the speed and position of the balls from both being specified to arbitrary precision. If one or the other is sufficiently inaccurate, the alarm clock will fail, and hibernation will turn into eternal rest.

One might imagine other alarm clocks that could forever remain above the quantum limit and might even be integrated into the organism itself. Nevertheless, no one has yet come up with a specific mechanism that could reliably wake an organism while consuming finite energy.

The Eternal Recurrence of the Same

The third and most general doubt about the long-term viability of intelligent life involves fundamental limitations on computation. Computer scientists once thought it was impossible to compute without expending a certain minimum amount of energy per operation, an amount that is directly proportional to the temperature of the computer. Then, in the early 1980s, researchers realized that certain physical processes, such as quantum effects or the random Brownian motion of a particle in a fluid, could serve as the basis for a lossless computer. Such

---

The Worst of All Possible Universes

Among all the scenarios for an eternally expanding universe, the one dominated by the so-called cosmological constant is the bleakest. Not only is it unambiguous that life cannot survive eternally in such a universe, but the quality of life will quickly deteriorate as well. So if recent observations that the expansion is accelerating are borne out, we could face a grim future. Cosmic expansion carries objects away from one another unless they are bound together by gravity or another force. In our case, the Milky Way is part of a larger cluster of galaxies. About 10 million light-years across, this cluster remains a cohesive whole, whereas galaxies beyond it are whisked away as intergalactic space expands. The relative velocity of these distant galaxies is proportional to their distance. Beyond a certain distance called the horizon, the velocity exceeds the speed of light (which is allowed in the general theory of relativity because the velocity is imparted by the expansion of space itself). We can see no farther. If the universe has a cosmological constant with a positive value, as the observations suggest, the expansion is accelerating: galaxies are beginning to move apart ever more rapidly. Their velocity is still proportional to their distance, but the constant of proportionality remains constant rather than decreasing with time, as it does if the universe decelerates. Consequently, galaxies that are now beyond our horizon will forever remain out of sight. Even the galaxies we can currently see—except for those in the local cluster—will eventually attain the speed of light and vanish from view. The acceleration, which resembles inflation in the very early universe, began when the cosmos was about half its present age. The disappearance of distant galaxies will be gradual. Their light will stretch out until it becomes undetectable. Over time, the amount of matter we can see will decrease, and the number of worlds our starships can reach will diminish. Within two trillion years, well before the last stars in the universe die, all objects outside our own cluster of galaxies will no longer be observable or accessible. There will be no new worlds to conquer, literally. We will truly be alone in the universe.

—L.M.K. and G.D.S.

computers could operate with an arbitrarily small amount of energy. To use less, they simply slow down—a trade-off that eternal organisms may be able to make. There are only two conditions. First, they must remain in thermal equilibrium with their environment. Second, they must never discard information. If they did, the computation would become irreversible, and thermodynamically an irreversible process must dissipate energy.

Unhappily, those conditions become insurmountable in an expanding universe. As cosmic expansion dilutes and stretches the wavelength of light, organisms become unable to emit or absorb the radiation they would need to establish thermal equilibrium with their surroundings. And with a finite amount of material at their disposal, and hence a finite memory, they would eventually have to forget an old thought in order to have a new one. What kind of perpetual existence could such organisms have, even in principle? They could collect only a finite number of particles and a finite amount of information. Those particles and bits could be configured in only a finite number of ways. Because thoughts are the reorganization of information, finite information implies a finite number of thoughts. All organisms would ever do is relive the past, having the same thoughts over and over again. Eternity would become a prison, rather than an endlessly receding horizon of creativity and exploration. It might be nirvana, but would it be living?

It is only fair to point out that Dyson has not given up. In his correspondence with us, he has suggested that life can avoid the quantum constraints on energy and information by, for example, growing in size or using different types of memory. As he puts it, the question is whether life is "analog" or "digital"— that is, whether continuum physics or quantum physics sets its limits. We believe that over the long haul life is digital.

Is there any other hope for eternal life? Quantum mechanics, which we argue puts such unbending limits on life, might come to its rescue in another guise. For example, if the quantum mechanics of gravity allows the existence of stable wormholes, lifeforms might circumvent the barriers erected by the speed of light, visit parts of the universe that are otherwise inaccessible, and collect infinite amounts of energy and information. Or perhaps they could construct "baby" universes and send themselves, or at least a set of instructions to reconstitute themselves, through to the baby universe. In that way, life could carry on.

The ultimate limits on life will in any case become significant only on timescales that are truly cosmic. Still, for some it may seem disturbing that life, certainly in its physical incarnation, must come to an end. But to us, it is remarkable that even with our limited knowledge, we can draw conclusions about such grand issues. Perhaps being cognizant of our fascinating universe and our destiny within it is a greater gift than being able to inhabit it forever.

r some fits and starts, the
ry seemed to best accommoda
growing observations. Tiny
tuations found throughout t
ic microwave background ech
erse's birth were one of t
y hallmark proofs of the bi
rs soon followed. Most mode
ological thought frames its
big bang theory. When its

Conclusion

U ltimately, the stars in the night sky are responsible for the musing and wondering that goes hand in hand with cosmological investigation. It is the stars. Their mystery has called to the human imagination, and in answering the call, the consciousness of the family of man has expanded itself into the consideration of the very materials and the very dynamics that frame all that is.

None of the theories or ideas presented here pretend to be the ultimate answer to the questions of where the universe came from, its exact nature or its ultimate fate. None of them account for observations in a way that would rule out all other possibilities and propel one idea alone from the status of a mere theory into that of natural law. Perhaps whole new worlds of thoughts unimagined now will become the commonplace theories of cosmology in the next century, or the one after. We just don't know.

We do anticipate that our technology will improve, though, and with it we expect the quality and depth of our observations of the universe to improve. As more pieces of the puzzle fall into place we expect our inquisition of the stars to become more pointed. We expect forays into fruitless lines of thought, and trust the dynamics of science to winnow out the false from the true. Yet with every small step we take, we expect to come closer to the truth of how the universe formed, evolved and developed, and what it means to us.

In the space of less than 500 years, the study of the universe and its parts has taken humanity from a geocentric, planet-bound consciousness to a knowledge of our place in the solar system, to the revelation that our entire galaxy is but the equivalent of a grain of sand on a vast beach of galaxies that fills the universe. It has opened us to the concept of infinity, and given us an appreciation of the scale that encompasses all that is.

Could it be that, on some level, quantum cosmology is right? Could trillions of worlds not unlike our own co-exist with ours in the space-time continuum alongside trillions of probable worlds, each safely couched within its own universe? Could the theory of inflation be a true reflection of the way nature weaves ever-evolving universes together, like pearls on a string, each one part of the greater whole but also unique unto itself? Could there truly be a theory of everything so simple and so elegant that its basic concepts could be understood by a child?

These questions may never be answered in our lifetimes. They may never be answered at all. But merely raising them adds a beauty and a subtlety to human thought. As we study the cosmos and ask questions, we exercise the human prerogative to wonder, to seek, and, hopefully, to find. It is part of the human heritage of belonging to the universe.

r some fits and starts, the
ry seemed to best accommodat
growing observations. Tiny t
tuations found throughout th
ic microwave background echo
erse,s birth were one of th
y hallmark proofs of the big
rs soon followed. Most moder
ological thought frames itse
big bang theory. When its

Index

PHOTO CREDITS: Page 42, Jared Schneiderman Design. Page 51: Alfred T. Kamajian. Page 53: Courtesy of Leven Wadley, Columbia University. Page 61: Dmitri Krasny. Page 64: Dmitri Krasny. Page 81: Don Foley. Source: Robert R. Caldwell, Dartmouth College, and Paul J. Steinhardt. Page 110: George Retseck. Page 119: Don Dixon and George Musser. Page 122: Don Dixon and George Musser.